995
88

FLOWERS OF GUATEMALA

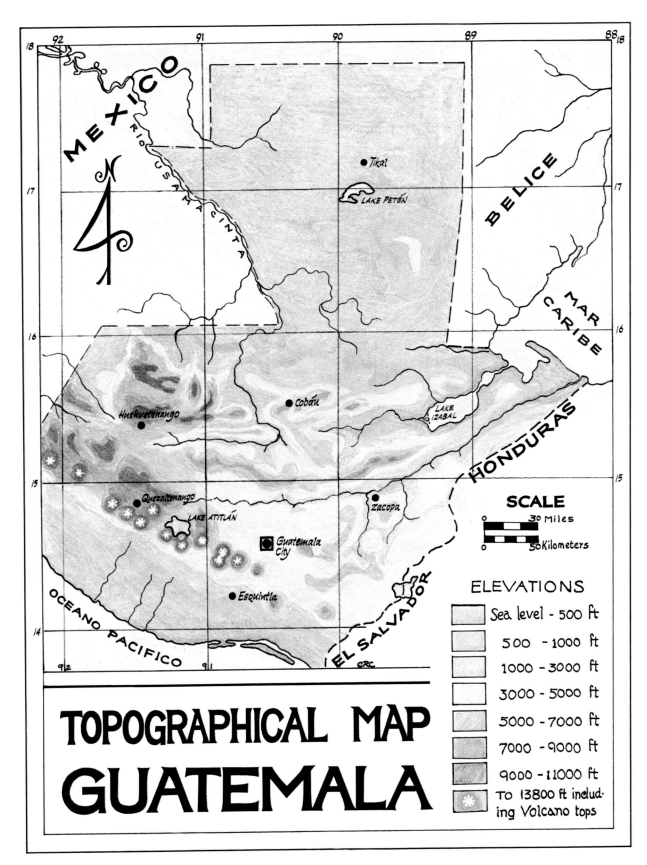

TOPOGRAPHICAL MAP OF GUATEMALA

FLOWERS OF GUATEMALA

by Carol Rogers Chickering

Foreword by Julian A. Steyermark

University of Oklahoma Press : Norman

Also by Carol Rogers Chickering

Flowers of Lake Tahoe (1970)

Library of Congress Cataloging in Publication Data

Chickering, Carol
  Flowers of Guatemala.

  1. Botany—Guatemala. I. Title.
QK218.C54   582'.13'097281   72-9278
ISBN 0-8061-1081-3

Copyright 1973 by the University of Oklahoma Press, Publishing Division of the University. Composed and printed at Norman, Oklahoma, U.S.A., by the University of Oklahoma Press. First edition.

This book is affectionately dedicated to my three children, Patsy, Joan, and Howard.

I hope that it will give them as many hours of pleasure as its preparation has given me.

Foreword

Julian A. Steyermark

Guatemala, land of the quetzal bird, is a botanical paradise. Whoever visits this Latin-American country finds a land admirably suited to plant adventurer and plant fancier alike. Guatemala has something to satisfy everyone, not only the experienced botanist and horticulturist but also the plain gardener and plant lover.

Guatemala is able to satisfy every taste because of its tremendous diversity of landscape, climate, altitude, and soils (all within such a small area). Here are mountains and volcanoes by the score; picturesque alpine lakes and meadows; cool, temperate oak and pine, fir, or cupressus forests contrasting with giant tropical liana-hung trees of rain forest, and moisture-laden, moss-festooned cloud forests; while hot deserts and steaming jungles contrast with cold, windswept summits of volcanoes and high mountains. Here are topographical features ranging from steep canyons, caves, evergladelike country, sulfur fumaroles and springs, waterfalls, clear spring-fed mountain streams to dark-brown waters of lowland areas.

Here is a land where north meets south, where such familiar northern trees as fir (*Abies*), juniper (*Juniperus*), bald cypress (*Taxodium*), hop hornbeam (*Ostrya*), American hornbeam (*Carpinus*), sweet gum (*Liquidambar*), and box elder (*Acer negundo*) are found at their southernmost known limits of geographical distribution, while such South American genera as *Greigia*, *Crumenaria*, and *Aphanactis*, and such species as *Plantago tubulosa*, and *Werneria nubigena* attain their

[1] Excerpt from an article which originally appeared in the *Journal of the New York Botanical Garden*, January, 1965.

vii

northernmost outposts. Here in Guatemala the familiar partridgeberry (*Mitchella repens*) has found an isolated refuge in a remote cloud forest of the Sierra de las Minas as its southernmost known limits. Other familiar eastern and southern United States plants, such as supplejack (*Berchemia scandens*), Virginia creeper (*Parthenocissus quinquefolia*), and Carolina or yellow jessamine (*Gelsemium sempervirens*), are at their southern points of range. Travelers familiar with western plants will find in Guatemala such genera of the Rocky Mountains as *Calcochortus*, *Cupressus*, wallflower (*Erysimum*), cinquefoil (*Potentilla*), lupine, arbutus, paintbrush (*Castilleja*), beard tongue (*Penstemon*), and snowberry (*Symphoricarpos*). Here the enchanting rain forests of the Río Dulce harbor a rich variety of tropical vegetation similar to that of many South American forests. In the north mahogany and chicle forests hold supreme with a wealth of tropical flora.

Thus from a land so diversified—in altitude, ranging from sea level to nearly fourteen thousand feet on the top of Volcano Tajumulco; in climate, with rainfall from six to two hundred inches annually, and temperatures from 100 degrees in the Motagua Desert to 16 degrees on the higher mountains; in soils, derived from rocks varying from sedimentary (limestone, shale, and sandstone), metamorphic (marble, serpentine, gneiss, and mica schists), to igneous (granite, diorite, pumice, and volcanic ash); and in topographic conditions, embracing plateaus, volcanoes, mountain slopes, lakes, valleys, and plains—emerges a kaleidoscope of infinite plant variety and contrast.

Guatemala harbors within its confines geologically ancient terrain. A part of the old core of Central America, untouched by glaciation and not covered by marine waters since remote times, is preserved along an east-west mountain axis extending from the Sierra de las Minas to the Sierra de los Cuchumatanes. In this area, available for plant settlement and occupation since late Cretaceous times, are found unusual kinds of trees, shrubs, vines, and herbaceous plants of great rarity and interest.

Preface

Many North American travelers in Guatemala are surprised to see familiar flowers along the roadsides, perhaps unexpected in a country in a tropical latitude. Yet many flowers cultivated today in gardens throughout the Temperate Zone are descendants of Guatemalan plants, whose seeds were collected and taken to Europe. This is true of herbaceous plants, as well as of flowering and shade trees, vines, shrubs, bulbs, and several edible plants. New World specimens were carried from Europe even to the Orient and later back to the New World. From the earliest days of Spanish occupation ships from the Netherlands, England, France, and Spain have never ceased taking plants back and forth, adding to the flora on both sides of two oceans.

For centuries the English especially have been great plant collectors and distributors between the hemispheres. Among the collectors were Captain James Cook and Captain William Bligh, of the H. M. S. *Bounty*. Maps were vague then, and it is difficult today to determine exactly where the original species of many genera came from, but the general areas are well known. Horticulturists the world over have worked slowly and patiently for decades to make the original Guatemalan plants hardier, brighter, and larger for today's gardeners. This is especially true of zinnias, dahlias, marigolds, and fuchsias.

High-flying birds have had the privilege of looking down upon this world of ours from great heights and seeing the vast panorama of oceans and rivers, of green and brown land masses, revealed in relief by shadows cast by the sun or moon. Today we can share this privilege: to fly down the length of our hemisphere in a modern plane is an unforgettable experience. We can not only comprehend the terrain over which we fly but also understand the geography and geology of Central America and the reasons why Guatemala is not all tropical, as might be expected from its nearness to the equator.

The flight from the north into Guatemala is thrilling and beautiful. Sometimes the visitor comes in over the Pacific shoreline of Mexico, but more often over the Gulf of Mexico and across Petén, the northern section of Guatemala, encompassing flat, primeval rain forests that follow the course of the great Río Usumacinta. Suddenly the flat expanse of green forest ends in massive mountain ranges piled east to west across the country from coast to coast. Along the Pacific Coast side of the airplane the startlingly perfect cones of volcanoes can be seen. There are thirty-three volcanoes, twelve of which are conspicuous and are over ten thousand feet high. All of them lie in an area about 150 miles long and within 50 miles of the Pacific Ocean. Only three are still active: Santa María Quetzaltenango, near Quetzaltenango; Pacaya, near Lake Atitlán; and Fuego, near Antigua, the last of which has had a small flag of smoke since last erupting in 1962. Nestled among these volcanoes is the gloriously blue Lake Atitlán. With the setting sun behind them, if the clouds "are not drawn about their shoulders as a shawl," they are a sight never to be forgotten.

Guatemala is a small country that at first seems easy to know well in a short time. But everything takes a while to be arranged. Trips from one area to another take longer than anticipated, for the mileage is misleading. Also, there are so many different kinds of regions that it would take a lifetime to investigate all of them. In my ten visits to Guatemala, however, I found it possible to acquire some knowledge of all the areas.

The country is about the size of New York State— forty-eight thousand square miles—and about one-eighth the size of Mexico. In but few regions is Guatemala more than two hundred miles from east to west. It rises from the sea to a maximum of fourteen thousand feet. The mountain ranges are part of the Cordilleras, which form the link between the northern and southern continents, bridging the gap between the Rocky Mountains and the Andes. This region has been peculiarly undisturbed geologically throughout the ages, and the land is thus extremely old as compared with most other regions of the Western Hemisphere. The highest range, forming a large plateau over ten thousand feet high, is called Altos Cuchumatanes; farther east is another range called Sierra de las Minas. The longest and largest rivers have their headwaters in these heights and flow hundreds of miles north to the Gulf of Mexico or east to the Caribbean.

With these essentials of the geography in mind, and with the picture as seen from the air—the little white towns tucked into the folds of the hills, the famous "vertical cornfields" on the mountainsides, and the deep-green forest-clad peaks of the volcanoes—we can begin to understand how Guatemala differs from other countries.

To paint the portraits of the flowers for this collection I made five trips to Guatemala at different seasons. Many plants bloom at the end of the dry season, in February and March; others, soon after the rains start in June; and a great many, as the rainy season concludes the year's cycle, October and November.

I explored the country with several friends, including Wilson Popenoe, who has written notes about some of the plants (indicated by the initials W. P. after the text accompanying the plates). We collected big jars of flowers and pressed many specimens for future identification. The text accompanying each plate indicates the botanical belt or belts in which the plant occurs and the place and time of year in which each specimen was collected.

I made the paintings directly from the live specimens within a day or two, while their colors were fresh. When the flowers are painted (which means that one must resist the temptation to employ the wonders of modern photography), the individual flowers of a spray can be portrayed distinctly and in detail. For identification of plants hitherto unknown, it is the only way. Painted specimens may look sparse at first glance, it is true, but the important, uncluttered flowers and stages of development can be shown.

Dr. Popenoe was responsible for kindling my enthusiasm to share my knowledge of the flora of Guatemala, much of which I learned from him. It is mainly thanks to him that I learned many interesting details about the plants on these delightful trips. He likes to be considered a "plantsman," and neither of us claims to be a botanist. For many years he traveled about Guatemala in his capacity as plant collector for the United States Department of Agriculture. During that time he brought to the United States the budwood of many avocados. Later he was associated with the United Fruit Company, experimenting with bananas and other fruits and crops. He is still assisting in experiments with Temperate Zone fruits in the tropical highlands. His famous home in Antigua, which he opens to the public, is a restored Spanish colonial residence two hundred years old. It is the subject of a book by Louis Adamic, *The House in Antigua.*

Among the very few authorities on the flora of this particular area is Julian A. Steyermark, of the Instituto Botanico, Caracas, Venezuela. He is co-author with Paul C. Standley of the *Flora of Guatemala,* many volumes of which have been published by the Field Museum of Natural History of Chicago. Steyermark collected plants throughout Guatemala on two major expeditions related to work on this project. That section of their Fieldhouse Series is as yet incomplete but is nevertheless the only authoritative descriptive writing published to date. Steyermark has also written for scientific and popular publications and is the author of several books, the latest being the *Flora of Missouri* (1963). He generously offered the use of a climatic map and gave me permission to use data from an article by him and Standley which appeared in *Plants and Plant Sciences in Latin America* (edited by Frans Verdoorn, 1949), as well as a portion of a feature article, presented as the Foreword to this book. I am greatly indebted to him and take pleasure in presenting this example of his contagious enthusiasm. The plants he lists will be significant to anyone with sufficient botanical knowledge to understand the amazing diversity of flora to be found in Guatemala.

Another of the few men who know the Guatemala flora is Antonio Molina, of the Escuela Agrícola Panamericana, Tegucigalpa, Honduras. He has given his time to check the identification of most of the flowers painted for this book.

I thank Elizabeth McClintock, of the California Academy of Sciences, San Francisco, and J. Dennis Breedlove and his wife, Vera, for assistance in collecting data on our material. I am grateful to the late James Kempton, who read this manuscript several times as it grew, and to Alan J. Galloway, of the California Academy of Science, for assistance with the final revision.

To these botanists and friends, and to my family, I acknowledge my gratitude, for without them the book could not have been produced.

CAROL ROGERS CHICKERING

Woodside, California
August, 1972

Contents

Color Plates

The paintings appear in alphabetical order by genus and species.

Maps

FLOWERS
OF GUATEMALA

Introduction—LOOKING FOR PLANTS

Guatemala can be divided into twelve climate belts, each with an essentially characteristic community of plants. These belts are the result not of latitude but of altitude. It is no wonder that Steyermark speaks of Guatemala with such boundless enthusiasm, for within its borders there are more than eight thousand known native species of plants. Although some of them can readily be seen by the roads and in populated centers, others must be sought out, in the right season, in restricted or isolated habitats.

The collection in this book is a combination of the common and the rare. Most of the plants are native to Guatemala; two or three were introduced a long time ago but now are an everyday sight. The months of bloom differ widely. Many plants flower with the warming weather and sunshine of the early months of the year, others flower when the rains start in May and June, and a great many bloom in the autumn after the summer rains are well advanced.

The main thoroughfares of the country have been surfaced, but most of the byways are still dirt roads. The very old roads, built up and reinforced against erosion with handlaid and fitted cobblestones, are rather narrow and steep, but a good local driver never has trouble. Little has changed since the turn of the century, except that there are now small buses (which often break down), and the Inter-American Highway extends the length of the country, though portions are not yet paved.

Distances are measured by time, not by miles, a fact which one should consider when asking for advice. Bridgeless rivers in the rainy season—June through October—can be very wild indeed. In the dry season a fine dust permeates the car and everything in it. All this is part of life in Guatemala. When it threatens to be annoying, one remembers that if the roads are ever perfected the uncommercialized scenery and peaceful atmosphere, so rare and precious today, and among the greatest attractions of Guatemala, will eventually be destroyed.

Botanical Climate Belts of Guatemala

The Twelve Belts

Petén (Belt 1)

For the average North American or European tourist, Petén is unquestionably the most dramatic department of Guatemala, because it offers the greatest contrast to what he knows at home. In this huge, flat rain forest, roughly 150 square miles, there are very few settlements and even fewer roads, and those passable only by Jeeps. The scattered towns would be isolated but for the local airline's sturdy little planes, which shuttle back and forth with supplies, chickens, pigs, mail, and drinking water. The airline also provides regular tourist service to Tikal, and the passengers are packed right in with the freight.

Tikal is a huge archaeological site, still half-hidden by forest. In 1956 reconstruction was begun under the direction of the Museum of the University of Pennsylvania. The project was turned over to the Guatemalan government in 1968, when the contract expired. By then, however, the university archaeologists had determined that the city was established at least one hundred years before the birth of Christ, centuries earlier than any other known Mayan settlement. The latest date established by artifacts is about A.D. 900. Thus Tikal prospered for at least one thousand years.

Tikal is unique in other ways. It is almost incredible that one can fly, in two hours, 150 miles from twentieth-century civilization and find a comfortable place in which to live and study archaeology and history two thousand years old and also virtually all aspects of the natural sciences.

The buildings housing the staff and visitors are constructed of plank or plaster walls with screened ceilings and windows. The roofs, like those of the natives' homes throughout the country are built on a separate frame of peeled logs secured with lashings of vines and thatched with huge, round palm leaves, accordion-pleated, about one foot thick. This structure does not rest on the walls; the frame supports it two or three feet above them.

There are a few good books about the ancient Maya civilization. One especially readable book for the layman who wants to understand the basic history and significance of the pyramids and their treasures is *The Rise and Fall of Maya Civilization*, by J. Eric S. Thompson. The story of Tikal is eloquently reconstructed in the pages of Thompson's book.

The flora of the region is utterly entrancing. William R. Brigham wrote in *Guatemala: Land of the Quetzal* (1887):

> Tropical vegetation cannot well be described. The real trouble that meets the novice on the threshold of the tropics is the utter inadequacy of the English language to express the variety and luxuriance he sees in the vegetable world. Even in color his vocabulary fails him and he must include in the name of "green" so many distinct tints that he often fails to try.

Besides the "greens" I found texture, shape, and sound. The foliage is of every imaginable shape and size. There is no sense that it is "impenetrable"—quite the reverse. The growth is open, except where cut over. Since the sun cannot reach through, the flowers are few. The sounds are those of tiny birds and, at dusk, raucous flocks of parrots, of hoots and squeals of monkeys, and of gentle winds a hundred feet above, in the tops of the tallest trees.

The trees are tall and straight, reaching up toward the sun. The trunks are whitened with a smooth lichen, making the trees difficult to differentiate. But no one can fail to recognize the huge philodendrons and monsteras hanging on old unexcavated walls or from branches of trees. The trees seem looped together high above with interwoven lianas, including air roots and the strangler fig (matapalo). This parasite starts

10

as a small vine on the trunk of a tree, but eventually, girdling the tree, it finally takes its place. Sometimes the ground at our feet is bright blue where blossoms of *Petrea volubilis* (Plate 30) have bloomed and fallen. In cleared areas free of trees are colorful pink stands of tiny begonia flowers. Bromeliads of parrot colors and yellow sprays of orchids cover horizontal branches of old trees. Giant ceibas are outstanding because of the enormous spread of their visible "buttressed" roots, and often the path is lined with fallen ceiba blooms. The yellow blooms of the *Tabebuia donnell-smithii* (Plate 39) reach high toward the sky and are conspicuous from the air. The chicle (chewing-gum) trees (*Sapota achras*) are easily recognizable because of the scars slashed by workers on the trunks. Many palms of varying sizes and types of leaves make up a significant percentage of the undergrowth.

Ocean Shores (Belt 2)
Lake Izabal (Belt 3)
Low Savannas of Izabal and Petén (Belt 4)

These areas occupy the northeast corner of Guatemala, bordering the Gulf of Honduras and Belize (British Honduras). Belt 2 comprises the coastal strips of the country, both on the Caribbean and along the Pacific Coast. It is characterized by swampland and tropical heat. The vegetation is mainly composed of mangroves (*Rhizophora mangle*) and button trees (*Conocarpus erectus*), except on rare protuberances of bedrock, such as those enclosing the mouth of Río Dulce, where there are many types of plants.

The area around Lake Izabal, forming Belt 3, consists of slightly higher ground—alternate savannas and woodlands. Belt 4 encompasses the foothills of the Sierra de las Minas. This area is cooler, with high humidity and seasonal rains, as well as cloud precipitation.

These belts and the characteristic plants defining the interrelated plant communities are shown on the map and accompanying descriptions on pages 6 to 8. Not many tourists visit these regions. Most of those who make the journey to Lake Izabal are bound for Quirigua and its great Maya stelae.

Puerto Barrios, in Belt 3, is easily accessible by car on the good highway from Guatemala City. An interesting trip from Cobán to Lake Izabal and Río Dulce is described in the following account (these travelers, friends of mine, were fortunate—the railroad has since been discontinued):

> We left on a local bus before dawn. Several hours later we were in Tucurú, where a missionary couple invited us in for coffee. Most of the people we saw were girls waiting for the bus to Cobán with huge trays of calla lilies on their heads. Another bus landed us in Pancajché about noon. After a wait, a train arrived which looked like a fugitive from a museum, or a TV western. Its little engine was wood-burning and the two cars were scarred with burns, as were we by the time we arrived at Panzós. Every half hour we would stop to pick up wood piled along the track, and after the fire was built up, a boy ran through the cars with a bucket of water dowsing little fires. We followed the gorge of the Polochic into lush, tropical foliage and brilliant flowers. We "put up" in and with the primitive Guest House. Unexpectedly, the night was cold and there were no blankets.
>
> The next morning in a small boat we entered Lago Izabal after going through forests and seeing monkeys, parrots and alligators. The forests were jungle-like. Although we could not see from the river, we knew the land sloped up away from the lake and changed into open savannas. We chugged several hot hours through the lake to San Felipe. Monkeys chattered in the trees, green and blue parrots flocked across our bows, and we knew the water teemed with alligators and tarpon. The sun was low as we entered Río Dulce. There were huge rock cliffs with dense tropical foliage and vines, magnificent scenery, and the beauty of Río Dulce was outstanding.
>
> The next morning at dawn we went by launch to Puerto Barrios in the rain. We took the little train to Guatemala City through miles of banana plantations, passing through Bananera (the United Fruit Company Headquarters) and Quirigua, famous for its enormous stelae in beautiful condition. We had a very good lunch at the Railroad Hotel in Zacapa. We arrived in the City in time for dinner, hot and

tired, but we have long ago forgotten our discomforts as we think of the fabulous scenery.[1]

In *Guatemala: Land of the Quetzal*, William F. Brigham described a trip from the coast to Cobán, on the north side of Lake Izabal, one hundred years ago. He walked and occasionally rode a mule on the journey, which took many days. He was impressed by the variety of plant life he saw, by the birds he observed, and by the Indians he met.

This book contains no paintings representative of plants of the areas east of Zacapa, but many plants in Belt 3 are also seen in Belt 8, the Bocacosta, a region of fertile, semitropical *fincas*. Belt 4 is subject to seasonal rain rather than daily condensation, and many plants are unique to the area. Indeed, a whole book could be written about Belt 4.

The suspension bridges, constructed of liana and woven reeds, have been described and pictured in early books on the region, especially those of John Lloyd Stephens, illustrated with the famous drawings of Frederick Catherwood. Walking across them poses problems as I learned. Once I was faced with the alternative of negotiating a bridge over the high, turbulent white rapids of the Polochic or walking several miles back to where I had come from. The handrails were too far apart to reach both sides at the same time. When I stepped on the plank floor, it wobbled sidewise with a sickening lurch. For about an hour I sat on a rock and waited to see how someone else accomplished the feat. Finally a native girl came swinging by, and I watched carefully as the plank bounced to her quick, trotting walk, one foot in front of the other. So that was the secret! I took off behind her, almost trotting, and had no trouble at all.

Rain Forests of Baja Verapaz and Alta Verapas (Belt 5)

The departments of Baja Verapaz and Alta Verapaz offer the visitor scenes unique to Guatemala. One travels from Guatemala City (in Belt

[1] Letter to the author.

13

10), through the Motagua River valley (in Belt 6), and then into the Sierra de las Minas. This mountain range, the weather barrier at the southern border of Petén, is the background for another "rain forest" but in almost all aspects is unlike the rain forests of the north.

Belt 5, surrounding the Cobán and including the bewitching little towns of Tactic, Tamahú and Tucurú on the Río Polochic, is cool, with weather that always seems on the verge of breaking into a gentle, spring-like rain (unless it actually is raining). It is refreshing to see green grass, fruit-tree blossoms, and a great variety of ferns and shrubbery under all kinds of trees which one would never expect to live together in harmony. The roadsides are gay with vermilion lantanas and with terrestrial orchids—the scarlet *Epidendrum radicans* and the lavender *Bletia purpurea*.

One can fly to Cobán, but an automobile trip with a Guatemalan driver is a rewarding experience, especially if he is in no hurry. The round-trip drive from Guatemala City can be made in one long day (twelve hours with many stops) by way of El Rancho. I suggest keeping at least one day free to walk down the road from Tamahú and along the Río Polochic. I collected many flower specimens there.

The mountains rise only four to seven thousand feet and so are never really cold but are always damp. The forests of oaks, pines, and mixed evergreens are also the home of the Guatemalan national flower, the beautiful orchid called the white nun (monja blanca), botanically known as *Lycaste virginalis alba* (formerly called *L. skinneri*). The forests are the native habitat of many odontoglossums and other orchids well known to orchid fanciers (Plates 26 and 28).

Many European plants have been imported to Guatemala, and the gardens around the country homes of the Indians are always gay and bright. Among the plants that I was surprised to see was a Brazilian erythrina beside a magnolia grandiflora. Nearby was a stream of cool water flowing through a meadow with great clumps of what looked like iris. I was told that they were ginger lilies, which to me meant crinum

14

lilies. Later, after crawling through a fence and down a mud bank of the stream, I found that the plants were the very delicate white ginger (*Hedychium coronarium*), commonly used in Hawaiian leis. There was no end of surprises in those little green valleys in the hills—imported roses, geraniums, calla lilies, *Watsonia* and gladioli.

Farming is as efficient here as anywhere else in the country. Most of the landowners are of German descent, and the coffee and tea *fincas* are flourishing. The handsome girls of the region walk in the typical swinging gait, their short, white huipils free to fly out in back, and their long braids are caught in gay six-inch-wide bows. The people market their specialty of silver jewelry and, of course, all kinds of flowers grown in the area, as well as vegetables, beans, and corn.

At one time a great many little palm trees were collected in the surrounding hills to be sent to Florida for experimental studies. These were called *Nepenthe bella*, although, according to Steyermark, they are actually *Chamaedorea elegans*. Unlike other small palms, which have been cultivated the world over as potted plants, these never grow more than three feet high and remain neat and "dainty." Today they are extremely rare, as are the *Lycaste* and other, more showy orchids. No limits have been placed on exporting these plants. If such limits are not imposed, the palms will probably become extinct.

On the flight home from a recent trip, the plane dropped down into the dry Motagua Valley through range after range of hills that grew browner and dryer as we lost altitude. There were many interesting specimens to pluck for my press—a pink *Bixa orellana* (Plate 4), a lovely *Plumeria rubra* (Plate 31), a budding *Cochlospermum vitifolium* (Plate 14). From Salamá an old road leads through the mountains back to Antigua through Rabinal and Xenocoj—a rough but safe road for a strong car.

INLAND DESERT
(Belt 6)

The typical region of the Motagua Valley includes the same point, El Rancho, that we reached when we turned north to go to Verapaz and to Salamá. Its character is formed by mountains on all sides, which cut off ocean breezes and therefore cloud condensation. The protection from the influence of sea winds also extends east and south, reaching down along the borders of Honduras and El Salvador. There is a regular rainy season, and the actual rainfall is much like that of Belt 10, but much of the soil is rocky or gravelly and so porous that it drains quickly. Therefore, the area is truly an inland desert.

It is easy to drive to Zacapa in a day, stopping to enjoy the primitive and naïve façades of some old churches, my favorite being that at San Christóbal. On a bright day in June, after the first of the seasonal rains, all the foliage on the trees and shrubs has been washed clean. The ground here and there is covered with masses of orange, white, and lemon yellow, the flowers of *Amoreuxia palmatifida* (Plate 2), of convolvulus, and of *Tribulus cistoides* (puncture vine), which is a relative of *Guaiacum sanctum* (lignum vitae, Plate 20).

Outstanding are the different-colored flowering trees among the organ cactus and chaparral. Along the road are planted royal poincianas, but the dry wash of the wide river bed is full of native trees: the brilliant blue-canopied lignum vitae, the feathery pink mimosas, and the plumerias, with their strangely glowing starlike white flowers bunched at the end of leafless twigs resembling cigars. There are white-flowering *Cordia alba* (Plate 15) that look like apple trees, bright-green-foliaged *Crescentia alata* (calabash trees), and everywhere yuccas and organ cacti. Many numbers and varieties of cacti grow in this belt.

The valley is oven-hot in June, and it was a relief to arrive in Zacapa. It, too, is hot and dusty, and its chief attraction is its lovely flowering trees —blue jacarandas and red poincianas, six-foot-high poinsettias in dooryards, and showers of bougainvillaea and *Petrea volubilis* (purple wreath).

The huge old wooden Railroad Hotel, conspicuous for its two high stories and enormous rusty roof, is cool and airy upstairs, with a huge porch shading the bedrooms on all sides. The furnishings are old, but the hotel was spotlessly clean when I visited it, and we were served an excellent dinner. Afterward we sat on the upper veranda and watched the little trains come through, the platform being actually a part of the building. It was a gay, friendly, and on the whole delightful evening.

The hotel was built about 1905. The railroad was a very ambitious undertaking, for it connects the ports on the Pacific Coast with Guatemala City and thence to Puerto Barrios on the Caribbean Gulf.

One part of Belt 6 that no one seems eager to see—or, at any rate, to revisit—is the dry, desertlike country between Esquipulas and Zacapa. The famous pilgrimages to Esquipulas appear to be the only reason to travel through it—except for the naturalist.

Of special interest is the desert area bordering on Honduras, an isolated region in which plants grow that are found nowhere else except in South America. Recently one of the most spectacular wild cattleya orchids, *Laelia digbyana*, oversized and ruffled as any hybrid could ever be, was found in the mountains near there.

PACIFIC PLAINS
(Belt 7)

This belt is clearly defined both on the map and on the ground. Its characteristics are unlike those of any other area of the country. That was made clear to us when we flew over Chiapas from Oaxaca, Mexico, through Tapachula. Flying low, we could see the whole Pacific Plains divided into squares, where bananas, papayas, and many other crops were growing. The area is roughly 50 miles wide and 150 miles long from the Mexican border to Escuintla. Tourists who travel from Guatemala City on the good but less-well-known lowland road pass between two beautiful volcanoes, Pacaya and Agua, and generally go to Quetzaltenango, turning off at Mazatenango.

The countryside is different from the other belts. The land is flat and low, rising from sea level to two thousand feet. One passes through a long stretch of what seems a wasteland, but off the road on one side are large coffee *fincas*, and on both sides grow other crops, such as sugar cane, grains, and fruits.

The town of Escuintla is strategically situated at a junction of roads to the Pacific and also east and west. It is a shipping center for tropical-fruit growers. It is also a convenient place to stop to buy gasoline or to have an excellent luncheon at Sarita's of fresh-water crayfish soup and ice cream.

Belt 7 has a distinct summer rainy season and a cool, damp cloud covering almost every night. The daytime heat and the damp nightly cooling are favorable to luxuriant growth. Hundreds of small, clear streams spill down from the volcanoes fringing the plains; and under gorgeous trees many settlements or small villages surround the streams along the road. The dooryards are framed with banana plants and mango, avocado, breadfruit, and papaya trees. The walls support colorful clumps of orchids, blue convolvuluses, orange trumpet vines, and magenta bougainvillaea. Along the road are tall grasses alternating with huge banks of *Caesalpinia pulcherrima* (Plate 5) and clumps of cannas (Plate 7).

Between these green oases are brown fields lying fallow; in them stand isolated trees that are conspicuous when blooming. Of course, the great ceiba, or silk-cotton tree, is always easy to spot, growing one hundred feet high, of large girth, and with a true canopy of huge branches in a horizontal position. Most of the ceibas are laden with garlands of epiphytes, showing brilliant bloom at different seasons—sprays of yellow oncidiums, purple cattleyas, and dagger-leaved bromeliads with scarlet, blue, and orange flowers. Occasionally the typical buttressed roots of the

Bombax family are visible on the *Ceiba pentandra*, which yields kapok. This tree grows wherever it is warm enough with sufficient moisture in the air.

There are pink sweet-pea trees (*Gliricidia sepium*, Plate 19). They have many names (madre de cacao is a common one) and as many uses. They serve as shade for young orchids. There is the small golden-yellow buttercup tree (*Cochlospermum vitifolium*, Plate 14). Here and there are orchards of *Bixa orellana* (achiote), which are pruned back as far as possible. The papaya trees look like caricatures of the royal palm. *Tabebuia donnell-smithii* (Plate 39) can be seen from the road. There is also a pink-flowering variety.

In this belt, which is thirty to fifty miles wide, many crops are raised for local, national, and foreign consumption. Coffee does not do well at lower altitudes. The leading crops are sugar cane, which is fifth in economic importance in the country, and corn, fourth in importance. Of export and domestic value are bananas, grown here and on the Atlantic Coast. Cacao, tobacco, and rice are also grown here, but none of these crops equals cotton in importance. Grains, fruits, and legumes are grown for local consumption.

In this amazingly fertile belt are many other crops, including indigo, rubber, vanilla, oranges, coconuts, pineapples, bamboo and lemon grass, citronella, and balsam. Here are beautiful great trees, among them the cashew nut (*Anacardium occidentale*), of the same family as the mango and the pistachio. The avocado (*Persea americana*) is of the Lauraceae family, as are the bay (producing the leaves used as seasoning) and the cinnamon. Guava (*Psidium guayaba*) and papaya are among the favorite fruits.

The Indians eat a great deal of fish, which helps fulfill their protein requirements. The diet of the Indians is relatively high in protein owing to the quantity of beans and corn, but animal protein is not readily available. Poultry is easy to raise, and pigs and lambs are also raised, but they are usually sold when very young and small—the Indians make little money from them and do not reserve any of the meat to feed to their families. Cattle are scarce and very small; feed is at a premium since its elements are among the basic requirements of the people.

East of the Pacific Plains the landscape tips up to what seems like a green wall. Indeed, this is truly so, for here are the bases of the series of volcanoes. The road leads halfway around Agua, then to the foot of Fuego, then to Atitlán and San Pedro. The volcanoes are green with plant life all the way to the peak of the highest one, twelve thousand feet high. Driving north through the village of Mazatenango, one takes the road for the pass through the volcanic wall to go to Quetzaltenango. Tourists seldom go beyond Mazatenango unless they are driving through to Tapachula, Mexico; but in this area, roughly thirty miles across, is some of the finest *finca* land in the country. With the largest volcanoes as a backdrop (both Tacaná and Tajumulco are over thirteen thousand feet high), it is an unbelievably gorgeous sight.

BOCACOSTA
(Belt 8)

Along the inland border of Belt 7 the ground suddenly ceases to be level and rises very abruptly, from two thousand to ten thousand feet. From below the lower part looks as though it were the forested base of the series of volcanoes, but actually the real forests are from seven thousand feet up. The elevation of Belt 8 is two thousand to four thousand feet. This region is the Bocacosta. It is the most fertile and most propitious area for the coffee *fincas*; the backbone of the country's economy is coffee exportation.

Very few tourists enter this belt, for there are few towns and no archaeological or historical sites; it is purely agricultural. There are literally square miles of coffee trees, some grown under the shade of *Grevillea robusta*, which have orange-colored flowers like big toothbrushes. There is a distinctive, pleasant aroma when the coffee trees are covered with their little white flowers. When the berries are ripe, the pickers are busy.

On the *fincas* other crops are usually grown that keep the people busy all year and make them almost self-supporting. Produce that is not needed goes to nearby towns on market day or perhaps even to Guatemala City.

This belt is not quite tropical, and yet the weather is never cold. Throughout the year the moisture from the ocean is dropped by the nightly cloud cover, and there is an annual rainy season.

A good dirt road leads around the south end of Lake Atitlán and wanders down to the lowland main road. An Indian driver who can speak with natives beside the road is an asset, for there are many crossroads and never a sign.

Some of my flower specimens came from this area. In depressions where lively little streams gurgle, there is an amazingly lush growth; verdant shrubbery is crisscrossed with a tangle of vines under beautiful trees. Big canopy trees grow here and there. There is such a variety of plants that a complete list would be impractical.

Of course, the *finca* is not only a crop-producing farm but also the home of the owner. The houses are mostly old-fashioned—revealing a mixture of nationalities and eras, depending on the owner's family origin. There are beautiful gardens, some old European in design, others equally attractive, with a few flowering trees as specimens on a lovely extensive lawn. Often there is a small lake, used as a reservoir for water supply from the precipitous streams and also in landscaping as a reflecting pool. There are many birds, and always the intensively Guatemalan touch of a backdrop of at least one blue volcano cone. In spite of the many people living close by, the atmosphere is quiet, restful, and leisurely. These lovely places are typical of all Central America, but other countries seldom have such comfortable weather. There are few places in the world more desirable from this point of view.

CLOUD FORESTS ON VOLCANOES
(Belt 9)

Higher on the Pacific slopes of the volcanoes, from 4,000 to about 7,000 feet, above the commercially busy Belt 8 of coffee *fincas*, is a cooler region. It is warm in the sun, but an almost nightly cloud cover acts as a year-round rainy season. This is a cloud forest of broad-leafed trees with many epiphytes. Moisture-loving plants of all sizes—tree ferns, oaks, alders, holly, silk-tassel bushes (*Garrya laurifolia*)—thrive, as do ironwood, elderberry, great lianas, mosses, and small ferns. The endemic curiosity is *Rojasianthe superba*.

This area is seldom visited by tourists because of the major effort required to ascend the mountains above the *fincas*. The only place in this fascinating belt easily accessible is the region around Fuentes Georginas, a hot sulfur spring reached by turning off the road between Mazatenango and Quetzaltenango. On my last visit there the road to the spring had been washed out, which made it necessary to walk five miles, but it was worthwhile.

There are many hot sulfur springs in Guatemala, but Fuentes Georginas is especially attractive because it has not been commercialized. There is a cement pool about twenty feet square and fifteen inches deep. Into one corner drops a little waterfall from a tiny, bright ice-water stream. Into another glides a velvety smooth, steaming-hot flow from the spring. We ate lunch sitting in this "bath," having entered the water where it is temperate and moving at will from hot to cold, much as the Japanese do.

The pool is surrounded by tree ferns with bromeliads and orchids on all big branches of the enormous trees overhead—very different from the wintery brown-dry countryside from which we had come. In fact, one of the major attractions of Guatemala is that within the space of a few hours a climate can be found to suit every mood. Mexico, it is true, has the same feature, but since it is eighteen times larger, much more traveling is required.

CENTRAL HIGHLANDS
(Belt 10)

The best starting point to see the flowering plants of the Central Highlands is Antigua. The city stands at five thousand feet elevation in Belt 10, the best climatic belt in which to live and to grow Temperate Zone plants. Most of the modern-day towns are at altitudes of three thousand to six thousand feet. The Quiché Mayas lived in this region long ago. Plenty of water and fertile, sunny fields are available for growing vegetables, fruits, and fodder. The climate of the different regions has such a profound effect on the people that one of the worst punishments in the old days was to send highland Indians to the Atlantic Coast to work in the "banana country," where they were likely to contract malaria.

Here in the hills the days are hot, but the nights are cool, the temperature occasionally dropping to 45 degrees. The clouds gather around the volcanoes nearly every day about 4:00 P.M., and the rains come almost every afternoon during the summer. The resulting humidity is good for all forms of life. On the upper slopes of the volcanoes, which are in the clouds each day and receive even more moisture, is more luxuriant plant growth, including evergreen conifer forests.

To visit the Central Highlands, one has a choice of four roads from Antigua, one out of each "corner." One is the paved Inter-American Highway to Guatemala City and on to the Motagua Valley and points north and east. The second road is a continuation of this one and goes northwest toward the other highland towns and Lake Atitlán, there rejoining the Inter-American Highway. It goes first through Chimaltenango, from which several difficult-to-find unpaved roads lead to the small towns San Juan Sacatepéquez, San Martín Jilotepeque, and

Xenacoj. In this area many varieties of European imported flowers are grown for the markets, such as *Watsonia*, roses, Easter lilies, carnations, gypsophilas and calla lilies.

The third route out of Antigua is an inconspicuous dirt road that starts behind the great Church of San Francisco and leads up to the little, grubby, friendly town of Santa María de Jesús on Agua, at an altitude of sixty-five hundred feet. At this point the road forks; the road to the right leads toward the cone of the volcano and ends some distance on the sheltered side of the mountain. The view of Antigua Valley is beautiful, and the air is fresh and clean. Continuing straight, the road drops into the next valley and winds down through a cypress-shaded slope; in all the shaded corners are lovely flowers, ferns, and flowering shrubs. Below the junction of this road and the paved highway from Guatemala City is the little town of Palín, famous for its great ceiba tree, under which is a perpetual little market, where colorfully dressed women sell even more colorful tropical fruits, such as pineapples, mangoes, and papayas, as well as vegetables.

Another way to the highlands is the road from the fourth corner of Antigua, through Ciudad Vieja and Alotenango, and on to Escuintla. This road is not recommended to everyone, just to nature lovers who want to see another kind of vegetation. The road eventually crosses the black-ash and sand runoff, ten to fifteen feet deep, from Fuego, which erupted in 1932 and again in 1962.

It has been interesting to watch the development of new plant life, which on the volcanic material has taken place more rapidly than many of us dared expect. The Highlands, many parts of which must have been in cultivation for more than two thousand years, have had their fertility renewed by these occasional volcanic deposits. Once, while driving on the Pacific side with a famous geologist, I pointed to the new cone, Santiaguito, on the slopes of Santa María near Quelzaltenango, which was then in active eruption, and told him that several feet of ash had been thrown over the area in 1910. I ended with the comment, "This is a new country." The geologist sagely replied, "Yes, maybe it is not finished yet."

There is a noticeable change in plant life in this kind of soil. In the little folds of the mountains are many shade-loving plants, including *Aphelandra schiediana* (Plate 3). There are many lush varieties of wandering Jew (Commelinaceae), and of unusual interest are the little blue *Achimenes longiflora* (Plate 1) growing on the precipitous black soil banks. Finally the road goes into the open and across the huge, rocky dry wash of the Río Guacalate, where plants that like dry heat, such as *Cassia* (Plate 8), *Wigandia* (Plate 46), madroña (indio desnudo), and many thorny acacias are to be found.

UPPER HIGHLANDS
(Belt 11)

From the fourth "corner" of Antigua on the way to Escuintla by way of Alotenango is an almost unknown road to the right, up the valley between Fuego and Acatenango into Belt 11. This belt ranges in elevation from two thousand to four thousand feet.

Here, on the high, protected side of the volcanoes, a different world of beauty and fantasy is found, one so delicate and sweet that it hardly seems real. The road is a good, reasonably wide one for a long way and gradually drifts up into the clouds from the dry, steep fields. The valleys, or folds, at the corners of the road are green and succulent with small, gaily colored flowers and flowering shrubs—five-foot-tall pink salvias (Plate 37); wild fuchsias (Plate 18); and malvas (Plate 27); streamers of "peas and beans," vines of tiny, jewel-like sweet peas; and such assortments of small plants as grow together in the grass (Plate 25). High above them, their heads literally in the mists, are the nodding, swaying flower heads of the fifteen-foot-tall tree dahlias (Plate 16) interwoven with tangled convolvulus vines of white, pink, blue, and pale lavender. A small tree with silvery-blue flowers and grayish leaves of the mallow family is *Robinsonella cordata* (Plate 35). In the deep pocket around which the road winds is the largest of the strange little-hand-flower tree, which has two other, awesome names. We argued and could not decide whether the present-day botanical name, *Chiranthodendron pentadactylon*, is more easily pronounced than the Aztec name, macpalxochiquahuitl (Plate 12).

On Agua, too, flowers bloom the year round. There are many plants that do not grow below, but most of them are species of well-known genera. Among them are a blue lupine and *Castilleja tenuifolia* (Plate 9), an unusual type of Indian paintbrush.

ALPINE HEIGHTS
(Belt 12)

The land of the volcanoes is unique. Just the memory of the skyline they present takes one back to another world: a world of peace, quiet, and serenity, unmarred by any of the elements disturbing the natural beauty of other countrysides, a world where one is conscious of the sweep of "the twelve clean winds of Heaven" stirring the great cypress trees, swaying gently among the vines and creepers that festoon the trees, and murmuring among the cool ferns and mosses at their base.

In the country around Quetzaltenango, over the mountainside to Zunil, or on the road to Huehuetenango the crisp air, the quiet peace, and the windswept trees are an experience not often found elsewhere today. Quetzaltenango is high and chilly all year round, but it is the center of an enterprising neighborhood whose inhabitants produce good-quality sheep and goat wool. The wool is processed and woven there, much of it retaining the natural colors of grays, browns, blacks, and white. In the hill villages along the road are many little Indian homes with a *corredor*, or porch, in front, on which the father is hard at work at his sewing machine. He makes jackets and trousers for the city markets and also stitches straws and grasses for sombreros.

Only a few small buses and a few flocks of sheep are seen on the roads during the day. This is fortunate, because the roads, though of excellent surface, are narrow and wind around the mountains often one or two thousand feet above the bottom of the canyon.

We made two trips from Huehuetenango; one to Chemal (the "Windy Place"), situated at the end of the road, from which a trail descends thousands of feet north to Todo Santos, the town in which Maud Oakes chose to live for two years because it is as remote as possible from invasion of "foreign" ideas. She wrote her well-known and delightful book *Beyond the Windy Place* (1951) after living there as observer, judge, doctor, nurse, and sanitation teacher. I can recommend no better book about the lives of the Indians of the region.

Roads skirt the mountains to Nebaj and Sacapulas. When the residents saw us emerging from the car in the Nebaj town plaza, they disappeared. It took much byplay with cameras and much giggling to entice several small boys from hiding. A few men were lounging beside their doorways, and it must have been an hour later when the women finally decided we were not dangerous and came to the central fountain, or *pila*, with their ollas on their heads for their evening measure of water. In Sacapulas, we were charmed by the silver coin medallions the women wore on ribbons around their necks.

Returning to the lower regions by way of Aguacatán, we spent the night wrapped in blankets from Momostenango, in a little park beside a sparkling bright stream that gushes forth from a parched hillside.

As we drove back to "civilization," we finally reached the well-publicized Chichicastenango, on the rim of Belt 10. The market there is extraordinarily picturesque. The Mayan Inn is also a rare experience. Somehow one can feel that, regardless of the countless tourists who have been there in the last forty years, the Indians have not lost their identity.

The goods sold in the market remain strictly for Indians, except for some of the cloth, which is manufactured on commercial looms in Quetzaltenango, Totonicapán, and Guatemala City for export.

Before returning to city life, we spent two days at Panajachel, on the shore of Lake Atitlán. It was pleasant to walk about in the lovely gardens and swim in that so-blue lake. We spent one day on a launch trip to Santiago Atitlán across the lake. The little girls asked for a few cents when we took their pictures. With the women in their red and white clothes and the characteristic *halozo* around their shining hair and the men in striped short pants, their market was quietly attractive.

Their church with its very old wooden statues was charming. One day the saints were wearing cotton bandana handkerchiefs around their necks and *rebozos* and were standing on an altar covered with oilcloth. They were delightful.

Color Plates

Achimenes longiflora (flor de peña)   Belt 11, Alotenango

Gesneriaceae (gesneria family)   June

In their preliminary *Plantas de El Salvador*, Standley and Calderón listed this lovely little plant as flor de peña (rock flower). It is an appropriate name, for the plant grows on rocky cliffs and embankments, especially where it is shady and moist. It is abundant along roadsides around Agua Volcano and in many other parts of Guatemala at intermediate elevations. It springs up in what might seem to be the most unlikely places, for example, on adobe and masonry walls in Antigua. Its flowering season is late summer and early autumn.

The plant is small, the flowers large and showy. It is a favorite potted plant in the homes of Guatemalan flower lovers, of whom there are about as many as there are people. While the variety is violet in color, as shown in the painting, there are intermediate shades all the way to white.

It is of sufficient horticultural interest to be described and figured in Liberty H. Bailey's *Standard Cyclopedia of Horticulture*, where it is stated that there are perhaps forty species of *Achimenes* native to tropical America. In North America they are considered "greenhouse herbs." They blossom in late spring and in summer. They also make satisfactory house plants, much like African violets. Propagation is by means of the rhizomes, though cuttings may be rooted. Visitors who travel through the Guatemalan Highlands in late summer should look for this lovely little flower.—W. P.

Amoreuxia palmatifida (zacyab—Yucatan)   Belt 6, Zacapa

Cochlospermaceae (cochlospermum family)   June

Amoreuxia palmatifida in bloom is an extraordinarily lovely sight. We were fortunate to see it in June, just before we reached Zacapa in Motagua Valley.

This plant is a ground cover, or vine, ready to sprout perennially from a large tuberous root and to spread rapidly over a large area, covering anything in the way. Then, even surpassing the attractive foliage, comes this bright-orange flower.

At first sight the flower resembles a nasturtium, but a closer look reveals the unusual arrangement of spots on four of the five petals and the tiered banks of twelve stamens with purple and orange anthers. This plant is in the same family at the buttercup tree (*Cochlospermum vitifolium*—Plate 14) and *Bixa orellana* (Plate 4).

We saw blankets of this lovely plant in full bloom on terrain so dismally dry that it did not seem possible that anything could grow. Alternating with these bright masses were lemon-yellow puncture vines (*Tribulus cistoides*) in full bloom. I do not know whether this species of *Amoreuxia* grows elsewhere in Guatemala, but it is a rare plant.

Aphelandra schiediana Belt 11, Alotenango
Acanthaceae (acanthus family) June

This family includes a goodly number of showy herbaceous plants or small shrubs. Of the genus *Aphelandra* itself, Bailey lists in his *Standard Cyclopedia of Horticulture* no less than nine species with the following comment: "Tropical American evergreen shrubs or tall herbs grown in hothouses for the fine foliage and very showy four-sided terminal spikes of red or yellow gaudy bracted flowers." Other interesting genera in the family Acanthaceae are *Jacobinia*, *Anchusea*, and Thunbergia (the last-named a native of the Old World).

Aphelandra schiediana shown here was growing along the roadside near Alotenango on the highway between Antigua and Escuintla.

Bixa orellana (achiote) Belt 7, Escuintla

Cochlospermaceae (cochlospermum family) October

This attractive plant was named for Francisco Orellano, who "discovered" the Amazon River. We found this specimen in the wild, beside the road to Cobán in the hills overlooking Salamá and the Motagua River Valley. It is also indigenous to Belt 6. It was looking very happy with plenty of sun and water. It is a very large shrub, full and green, and covered with delicate flowers. It continues to bloom as the seeds ripen. Those plants in orchards or used by man are cut back drastically when the seeds are taken.

The word bixa is South American vernacular for "tree," which means that the plant is well known. It is no wonder, for it is used for many purposes. The seed pods resemble old-fashioned chestnut burs, and when they are ripe the velvety linings are red. The seeds are pulled out, and the linings are made into a brilliant yellow dye. It is digestible and is used for food coloring in margarine and as a substitute for saffron. These uses alone would be sufficient to make it worthwhile harvesting the wild fruit, but the Indians also use the juice on their bodies as a mosquito repellant. The wood is whitish and is used like basswood; the resin is a gum arabic; the bark makes strong fibrous cordage and also has medicinal properties: it cures dysentery and prevents scarring from wounds. No wonder the little trees we saw near Escuintla are pruned back almost like grape-vines in autumn!

Caesalpinia pulcherrima (dwarf poinciana, gallito) Belt 7, Escuintla
Leguminosae (pea family)   June

In many parts of Guatemala, usually at elevations below two thousand feet, this tall, sparsely branched shrub can be seen along the roadsides. Its brilliant yellow-and-scarlet flowers, though only an inch or two in diameter, resemble those of the royal poinciana in form and are sure to catch the eye of the traveler. Along the banks of the Motagua River, in the region below Zacapa, the abundance of the shrub suggests that it is indigenous, but botanists are not certain that this is true, even though the species is native to many regions of the American tropics.

Preference is given to the common name dwarf poinciana because it is meaningful to English-speaking persons familiar with the flowering trees of the American tropics, including Hawaii. The common name in the Zacapa area is gallito (little rooster).

The leaves of this plant are pinnate, the flowers borne in large terminal racemes. In addition to the typical red-and-yellow form, there is a variant with pure-yellow flowers. The seed pods are about two inches long, flattened, with several seeds, which germinate readily when planted in almost any moist soil. This plant has been grown successfully in southern Florida, but it cannot withstand frost. As shown by its behavior in the Zacapa region, it tolerates poor soils and thrives in regions where the dry season is long and hot.

Calliandra houstoniana (cabella de angela, plumita)   Belt 10, Antigua

Mimosaceae (pea family)   March

There are said to be about 150 species of *Calliandra* scattered rather widely over the world. Sixteen are described by Standley and Steyermark as native to Guatemala. One is *C. quetzal*, named by Captain John Donnell-Smith for the national bird of Guatemala. It is a very rare species, like the bird itself, and so we had to content ourselves with *C. houstoniana*, which is common along roadsides in many parts of the country. It is among the legumes with long tufts of brightly colored stamens instead of the pealike flowers (the large family sometimes called Papilionaceae).

Members of the *Calliandra* genus are classified by some botanists in the group Mimosaceae, the mimosa subfamily. This genus and its congeners are attractive shrubs in bloom during a large part of the year. It is said that the root bark contains an alkaloid which produces death by systolic arrest of the heart, but we doubt that the Guatemalan Indians know this.

Other flowers growing in the same places along the road are milkweed (*Asclepias curassavica*), which looks much like lantana; *Tecoma stans* (Plate 41); a *Ceanothus* with inconspicuous flowers; and many species of mints with red, sapphire, and violet flowers.—W. P.

Canna edulis (cucuyu)   Belt 7, Escuintla

Cannaceae (canna family)   June

The inflorescence shown here was collected in a wet area along the roadside near
Mazatenango, a town about twelve hundred feet in altitude, on the Pacific side
of Guatemala. It is a very showy canna, though the flowers are much smaller
than those of the magnificent horticultural cannas grown today in gardens of
North America. From the wild cannas of tropical America, of which there are
several species, have been developed the magnificent, many-colored forms now
popular in gardens throughout the world. Standley and Steyermark record that
Canna indica is also indigenous to this region and to many other parts of Guatemala.

One of the interesting features of *C. edulis* is the fact that it is commonly
seen in the dooryards of the Indians' homes, where in all likelihood it has been
cultivated since pre-Columbian times. While probably native to the *tierra caliente*,
it is also often seen in gardens at the elevation of Antigua, five thousand feet or
higher. The roots of *C. edulis*, known as achira, are eaten in parts of South
America.

Most cannas are called platanillo in this part of Central America, though in
Guatemala the large-flowered cultivated forms are usually known as *cucuyus*.—
W. P.

Cassia indecora (retama)   Belt 10, Antigua

Leguminosae (pea family)   Summer

Most members of the genus *Cassia* are large shrubs, as is the one shown here; others have the undeniable shape of trees. All of them are green and fresh-looking throughout the year. They are especially lovely when they are covered with golden panicles of flowers, which seems to be most of the year.

All the many species grow in warm regions with a definite dry season (Belts 6, 7, and 10) and thrive under a hot sun. In the early months of the year they are often seen along the road with species of *Wigandia*, *Tithonia*, and *Acacia* in bloom.

The Leguminosae family is divided into three sections. All have true seed pods and pinnate leaves divided into leaflets—from three to a dozen or more. They fold down in hot sun so as not to burn or lose their moisture. The senna section includes the cassias and many popular flowering ornamentals, among them the shower trees of Hawaii. The pink shower tree is American. Another favorite is the bauhinia, also known as the orchid tree because of its delicate, lovely flowers. The senna section is also known as the Caesalpinioideae. It is distinguished from the other two by the shape of the flower.

The Papilionaceae section includes plants with sweet-pea, or butterfly, flowers. To this immense group belong such forage as clover and alfalfa and lupine and wisteria, as well as the genera *Gliricidia* and *Erythrina*, including Guatemala's two native species of the latter, *E. berteroana* and *E. poeppigiana*. All the varieties of peas and beans the world over belong to this group.

To the third section, the Mimosaceae, belong plants that have conspicuous stamens with petals either absent or insignificant. Mimosaceae includes the multitude of acacias; the tamarind; the monkeypod, or rain tree (*Pithecolobium dulce*); and Hawaii's kiawe (wili-wili); as well as *Calliandra houstoniana* shown in Plate 6.

Castilleja tenuifolia

(Indian paintbrush, painted cup) Belt 11, Agua Volcano

Scrophulariaceae (figwort family) November

Members of the *Castilleja* genus are known in the United States, especially west of the Rockies, by the name Indian paintbrush. I doubt that many people look at them carefully enough to realize that in one small area there may be as many as four or five species. Because they are generally the same size and color, the details of leaves and bracts are not noticed. However, there are many species, differing from one another in shape. The one shown here is unique—a spreading "clump" about eighteen inches high with perhaps a dozen or more spikes of the largest blooms I have ever seen. This specimen was found on the side of the Agua Volcano, in the sun, just below the cloud belt.

These plants are true parasites and are often listed as tree dwellers. They grow perennially from roots established in rotted wood—generally fallen trees so old that they are often buried by soil—and the plants appear to be growing in ordinary soil. Another surprise is that the most colorful parts of the plant are the bracts, by which the flower is almost hidden.

The genus *Castilleja* belongs to the family of which the snapdragon is the best-known member. All flowers in this family have "irregular corolla"; in other words, the petals are of differing shapes. I remember seeing two other species of *Castilleja* in Guatemala (one at Iximché), both of which are small. Each plant has two or three stalks, about ten inches high. The stem and pointed leaves are almost black, very stiff and dry, and all the parts of the flowers are also pointed. The specimens I saw were in a sun-baked field and yet resisted being burned.

It should be observed that plants incapable of storing water usually have very narrow or small leaves, which help them retain a crucial minimum to exist until the next wet season.

46 PLATE 9

Cattleya aurantiaca   Belt 10, San Martín Jilotepeque
Orchidaceae (orchid family)   Spring

On the Pacific side of Guatemala this lovely little species of *Cattleya* is abundant, growing high in the trees in Belts 7, 8, and 10. In spite of its small size, the color is outstanding. It grows naturally around Escuintla, and we found it in ravines north of Antigua, where one would think the climate too varied and dry. Tolerant of variations in climate, it is a very easy plant to cultivate.

The other Guatemalan *Cattleya* important to the world is *C. skinneri*, which has a proud reputation of being one of the oldest and most important members of its genera in hybridizing. Although in the natural state the flowers are small and unadorned, they have a lovely color, ranging from lilac to purple, and they bloom in a cluster that has earned the name candelaria (Candlemas) from local Indians. The plant is very hardy and generally has at least four flowers to a stem. For these reasons it has always been successful in hybridizing—ever since Skinner took it to England about 1830. Taste changes with generations, and now the great, ruffle-fringed, purple orchid no longer satisfies the market. This has given *C. aurantiaca* its chance to prove a strong parent and impart various shades of yellow through vermilion to its descendants, as amateurs join professionals in creating myriads of new combinations of plants.

Wherever there are bromeliads in old trees, one should watch for orchids, especially in early months of the year. Most of them have grasslike foliage, and the pseudobulbs are close to the roots. Orchids are not parasites; they perch on a tree, their roots clinging to debris piled into mats on old branches. The pseudobulbs are "tanks," supplying moisture and nourishment during the dry months of the year.

PLATE 10

Cestrum aurantiacum (herba del perro)   Belt 8, Moca

Solanaceae (nightshade family)   February

Species of *Cestrum* are quite common, the authorities tell us, from Veracruz, Mexico, south through Guatemala, growing in warm, humid areas. It is especially common in Belts 7 and 8. The specimen shown here was growing under huge trees on the slope of Atitlán Volcano in the coffee *finca* region, and was a large rambling shrub interlaced with small vines. The raceme consisted of many more flowers than are shown here.

Cestrum aurantiacum (which means "yellow") is a common plant, one of about a hundred species. It can now be found all over Europe, under glass, and in warm regions of the United States. It has not been reported poisonous to human beings although other species are; however, it is said to poison cattle and sheep. It has been reported to be used as a narcotic.

The best-known *Cestrum* is *C. nocturnum*, originally from the Valley of Mexico, which is grown for its famous evening fragrance. The flowers of *C. nocturnum* are greenish-white to pale yellow. The juice and fruit are used in the treatment of epilepsy.—W. P.

Chiranthodendron pentadactylon (mano de León,

little-hand-flower tree, little tiger's foot)   Belt 11, Fuego Volcano

Sterculiaceae (Cacao family)   February

Few Central American trees are more curious or fascinating than this one, the macpalxochiquahuitl of the ancient Aztecs. The names mano de León and little-tiger's foot may be easier to pronounce, but how much more picturesque is the Mexican name. The Aztecs liked compound words, hence macpal ("hand"), xochitl ("flower"), and quahuitl ("tree"). The name opens up the whole world of Aztec botany. Possibly nowhere else in the New World was such progress made toward a logical system of plant classification as in the Mexican highlands. Of course, the system was not based on what we would today call botanical characteristics, but it was highly objective.

The flower of this tree bears a striking resemblance to a human hand (the Guatemalan likens it to a tiger's foot).

Occasionally in an Indian's cornfield a solitary showy specimen of this plant will be seen—a large, spreading, thickly branched tree with broad, leathery leaves. Why was it left when the land was cleared? There may well be some religious significance attached to it. The Mexican botanist Maximino Martínez wrote that for many years a single specimen of the species grew in the neighborhood of Toluca, where it was venerated by the Indians. They believed that it was the only one of its kind in the world and that the gods had willed that none other should ever exist.

The most convenient places for the visitor to see this tree are in the mixed forests on the slopes of Agua or Acatenango volcanoes, at elevations between seven thousand and ten thousand feet. It blooms most of the year, and the blossoms can be collected from the ground. The tree is related to the flannelbush (*Fremontodendron californicum*) of Mexico and California.

Cobaea villosa (herba acorda, cup-and-saucer)   Belt 11, Agua Volcano
Polemoniaceae (phlox family)   June

This species of *Cobaea* is rarer than the well-known *C. scandens*. It is a startling sight, twined about and over nearby shrubs for it glows like sunshine. The flowers are well separated from the leaves but are set, cuplike, among its green sepals, a characteristic which has led to the common name cup-and-saucer vine. It is one of a small genus of the phlox family, most of which grow in rather humid subtropical mountain areas from North America to Brazil. The name of the genus commemorates Bernabé Cobo, a seventeenth-century Spanish Jesuit and naturalist of South America.

C. villosa has one interesting characteristic unique in this family: the rachis of the pinnate compound leaf terminates in a tendril. Almost all such leaves look like many pairs of leaves with one at the tip (see some of the pea-family foliage). This tendril twines itself onto anything near each leaf as it grows, instead of growing more stem first.

This particular specimen was found high up on the side of Agua Volcano, spread out over a small manzanilla (little apple) tree. As far as we know, it has not been introduced into modern gardens. It grows very fast after the first rains of the season and starts blooming while growing. We have not seen it in the autumn.—W. P.

Cochlospermum vitifolium (tecomaxochitl—trumpet

    flower—buttercup tree, yellow rose tree)   Belt 6, Motagua

Cochlospermaceae (cochlospermum family)   Winter

This is a showy little tree in the dry and dusty landscape of the winter months. The golden yellow is a very rich color that is especially striking since the flowers are in full bloom before the leaves open. It is commonly seen in Belts 6 to 8. The Latin name means "shell seed, grape-leafed." Its varying common names indicate that the plant is well known and well liked. Indeed, it is found from the West Indies to Hawaii and all up and down the central section of the Western Hemisphere.

    The Indians obtain saffron from the stamens and use the "cotton" around the seeds for stuffing pillows. The wood is soft and easily broken, and if the branches are placed in soil, they will grow. Hedges can be formed this way, but the wood is too soft to make them worthwhile. The bark, the only tough part of the plant, contains a fiber used for cordage. A "decoction of the wood or leaves is a popular cure for jaundice and the flowers are employed as a remedy for chest infections" (Standley and Steyermark, *Flora of Guatemala*). Another account says that the tree is used in treating kidney and liver diseases and to hasten childbirth.

Cordia alba (zazamil) Belt 10, Motagua

Boraginaceae (heliotrope family) Summer

In the summer months of daily rains, one of the lovely additions to the landscape of the Motagua Valley is the *Cordia alba,* covered with white blossoms. It is reminiscent of old apple trees. The round crown is a sheet of white and very showy. In his *Flora of Costa Rica* Standley has written: "It is one of the charactertistic small trees through most of the drier parts of the Pacific tierra caliente of Central America. It is common from Mexico to West Indies to Venezuela." This tree is also seen in Hawaii.

The wood is lightweight and tough and is used for carpentry. Leaves and flowers are used for relieving chest congestion; a decoction is used for inducing perspiration; charcoal from the wood is used to treat stomach-aches. In Oaxaca the fruit is employed in coagulating indigo, and in El Salvador the "viscid juice" is used as a glue to fasten wrappers of cigars.

Among the plant's close relatives is the *Bourreria huanita,* well known in Mexico and Guatemala. It resembles *C. alba* in that it has white flowers, but they are a little larger, have blue stamens, and bloom for a long season with an exquisite fragrance. Isquixochitl, the Nahuatl name for *B. huanita,* commands a singular regard from the Indian of today—probably a legacy from the Aztec religion. Owing to Aztec influence, the Mayas for a short time sacrificed human beings. It is said that the very fragrant blossoms were used in religious rites as wreaths for the youth or maiden about to be sacrificed. Although the actual sacrifice is not now practiced, of course, the accompanying rites and ceremonies are remembered.

PLATE 15

Dahlia maxoni (flor de Santa Catarina)   Belt 11, Fuego

Compositae (sunflower family)   November

Along roadsides and in the dooryards of the highlands, between elevations of thirty-five hundred and seventy-five hundred feet (Belts 9 to 11) *Dahlia maxoni* is a conspicuous plant during late autumn. Though the typical form is lavender-pink in color, as shown here, there is a pure-white variant, as well as two double-flowered ones of the same colors as the single-flowered species.

D. *maxoni* grows to a height of about fifteen feet. Its stems are almost woody, two to three inches in diameter, and hollow. They terminate in huge panicles of flowers, which remain in bloom many weeks.

Residents of the Temperate Zone will be particularly interested in this plant because of its relationship as an ancestor to those popular garden plants the herbaceous dahlias, of which many beautiful varieties have been produced by plant breeders.

Propagation, as practiced by the Indians, is by means of cuttings one or two feet long, taken from large stems which are not woody but are not too succulent. Since this plant requires a long, cool growing season and is killed back by the first frosts of autumn, it is not easily cultivated outdoors in most parts of the United States, except in San Francisco's Strybing Arboretum. L. H. Bailey notes, however, in *Hortus Second*, that it has bloomed in Southern California. *D. maxoni* was first described by that great lover of tropical American plants W. E. Safford, from material collected in northern Guatemala about 1906 by William R. Maxon, of the United States National Herbarium. What seems to be a very similar species, *D. imperialis*, has been described from Mexico.—W. P.

Dahlia popenovii (cocoxochitl) Belt 10, Antigua

Compositae (sunflower family) July

This specimen was an exceptionally lovely one, named by W. E. Safford many years ago. Blooming from July and August throughout the wet months along the roadsides in the Central Highlands, these gay little plants are usually a lighter color, and the petals are narrower than the specimen shown here. They bloom in patches like small gardens along the road between Antigua and Guatemala City, accompanied by ageratums, salvias (mints), members of the *Cuphea* and *Bidens* genera, pentstemons, and many others.

It is refreshing to realize that these lovely plants have not been "re-created" by man. They are just as they appeared long before the Spaniards arrived four hundred years ago and took seeds back to Spain for gardens there, and even before Montezuma of Mexico ordered the vast collection for his storied gardens. (Owing to his influence, the remarkably objective Aztec system of naming was initiated and hybridizing begun in this part of the world.)

Cecile H. Matschat has given us the Aztec name for *Dahlia*, cocoxochitl (meaning "cane flower") and the information that the multicolored garden dahlias had already been hybridized when the red-flowering species, named for President Benito Juaréz, was taken from Mexico to England in the early nineteenth century. *D. juarezii* is now considered the parent of the modern "cactus" dahlia, and *D. popenovii* the grandparent. This account is symbolic of what Guatemala has to offer the world today—unspoiled, unmanipulated nature that is rare indeed in our times.

Fuchsia splendens (Adelaide)   Belt 11, Fuego Volcano

Onagraceae (evening primrose family)   Summer

Fuchsias are shrubs and trees native to Mexico, South America and New Zealand, but in most parts of the world they are raised under glass or as house plants in the winter. They appear to thrive in Guatemala, for several species grow there, especially in Belts 9 and 11. I know of three small-leaved plants, one a vine, growing on a house, which must be very old and has been pruned much as wisteria is treated.

Fuchsia splendens is startlingly different from other species. It is a tall, leggy shrub with thinly scattered leaves and flowers. We found it on the road to Fuego Volcano and identified it at the time only by its dark-blue berry. In Standard Cyclopedia of Horticulture, Bailey described the flowers as "scarlet with small greenish petals." They resemble the flowers of the lily *Fritillaria imperialis*.

The primrose family includes a surprising group of plants. Nearly everyone knows the evening primrose (*Oenothera biennis*), and some know the farewell-to-spring (*Godetia amoena*), a well-established garden hybrid, as well as a popular wildflower of the western United States. *Zauschneria californica*, whose common name is California fuchsia, is a member of the same family. It is a low-growing, sun-loving brilliant orange-red flower, rather like a snapdragon, growing high in the dry rocks in the mountains.

Gliricidia sepium (madre de cacao, mata ratón)   Belt 7, Escuintla

Leguminosae (pea family)   March

Imagine a tree thirty feet high, covered with pink sweet peas—a lovely sight. As one can see in the painting here, the leaves do not appear until each butterfly-like flower falls. These trees are popular for shading young cacao and occasionally coffee trees in Belts 7 and 8. They are native to the warmer regions of Central America and to Colombia. They belong to the Papilionaceae branch of the pea family (as differentiated from the Mimosaceae and Caesalpiniaceae). Other familiar Leguminosae are acacia, bauhinia, and erythrina, as well as genista, lupinus, poinciana, and wisteria. In fact, they may be found all over the world, both wild and cultivated.

Specimens of *Gliricidia sepium* are not often seen in full maturity (when they tower above the surrounding scrub), since they are so useful that the Indians cannot afford to let them grow undisturbed. The Australian grevilleas have taken their place as crop shade, and thus they are no longer planted for that purpose.

Among other uses they make excellent fence posts. A branch two or three inches in diameter, driven into the ground and encircled with barbed wire, will take root and grow in diameter to enfold the wire and tighten it. New shoots that are put forth are trimmed off for new posts. A four-wire fence is hard to get through; the wires are so tight that they hum. These fences are excellent for pastures and even better for home enclosures. They will keep pigs and chickens in (or out). Moreover, the roots poison small vermin and thus keep the dooryard and home safer and cleaner.

Guiacum sanctum (guayacán, lignum vitae)   Belt 6, Zacapa

Zygophyllaceae (caltrope family)   Spring

The valley of the Río Motagua, from El Progreso to Gualan, is often spoken of as a desert. Cactus and thorny shrubs give this impression. The hard, stony soil, rather than lack of rainfall, is responsible, though the valley is extremely hot and dry during half of the year.

It is during the dry season that this region blooms forth with some of the loveliest flowering trees to be seen in all of Guatemala. The guayacán is perhaps the one that leaves the most lasting impression on the visitor. It is a low, round-topped tree, densely clothed with dark-green foliage—a beautiful tree even when it is not in bloom. From March into summer it is covered with bright-blue flowers, followed by small, orange-colored fruit.

This tree is by no means limited to the Motagua Valley. It grows in many parts of Guatemala, but always at low elevations. While Standley and Steyermark accept the name *Guiacum sanctum* assigned to it in botanical nomenclature (which, of course, has to be accepted on the grounds of priority), I dislike to abandon the name Guatemalense, which has been used a long time. Because of this name and the unsurpassed beauty of the tree, one might suggest that this tree be named the national tree of Guatemala.

The guayacán is also appropriately called lignum vitae, since it produces a wood used for "bushings or bushing blocks for the lining of stern tubes of propeller shafts of steamships. . . . Its great strength and tenacity, combined with self-lubricating properties due to the resin content, make it especially valuable for bearings under water" (Standley and Steyermark, *Flora of Guatemala*).—W. P.

Heliconia bibai (platanillo)　　　　　　　　　Belt 7, Escuintla

Musaceae (banana family)　　　　　　　　　　October

Standley and Steyermark (*Flora of Guatemala*) describe ten species of *Heliconia* in Guatemala. All of them are called platanillo, the Spanish diminutive of *platano* ("banana"), because of the general appearance of the plants—they do not bear edible fruit. The leaves are sometimes used in public markets to wrap small quantities of salt and other products.

According to authorities, the species shown here is abundant in warm parts of Guatemala. A stand of plants is a spectacular sight. Its handsome inflorescences are sometimes cut and used in decorating altars and homes. Standley and Steyermark make the following comments: "There are probably fifty species, natives of Tropical America, several occurring in other parts of Central America and the genus extends northward into Southern Mexico. The larger plants are somewhat like the banana in habit, the smallest ones more suggestive of the genus *Canna*." Of the two best-known larger plants one has inflorescences that hang down, while those of the other are erect, as shown here.

Platanillos are confined to the hot lowlands of Central America, never extending far upward on the mountain slopes. In some parts of the *tierra caliente* they constitute an important and conspicuous part of the undergrowth in the forest or of the coarse second-growth thickets, forming colonies of wide extent. The leaves are six to ten feet long on a stalk, or trunk, about fifteen feet high. The inflorescences are about three feet long. These huge stands of enormous plants grow close, and they are ringed knee-deep with discarded bracts and dry leaves.

Ipomoea carnea (campanilla)   Belt 10, Antigua

Convolvulaceae (morning-glory family)   March

To illustrate the morning-glory family, I chose this species, *Ipomoea carnea*, because it is particularly clean and fresh-looking in contrast to the many pale blues and lavenders of other species. The vine from which this specimen was taken covered many square feet of surrounding shrubs. It is capable of very energetic growth. It will grow high into trees and even cover small houses. It is common in Belts 10 and 11.

Members of Convolvulaceae grow wild in equable climates all around the globe, and a great many have been brought into the horticultural world. There are about twenty species of *Ipomoea* in North American gardens and green-houses. One species is especially well known—the common sweet potato, (*Ipomoea batatas*)—which originated in the tropics of the New World.

In the rain forests there is an especially beautiful species called bona nox (*Ipomoea mexicana* or *Calonyction aculeatum*), the fabled moonflower. The large white trumpet spreads out to six or eight inches in diameter, and the heart is a clear chartreuse. We were lucky to obtain some seed, and grew it as a light screen across the top of a greenhouse of orchids. The growth was unbelievable. Unfortunately it had to be eliminated, for it was a host to aphids.

The plants that go by the name convolvulus are differentiated by the details of the flower structure. This is also true of the species of the genus *Quamoclit*.

Jacobinia umbrosa (monte de oro)   Belt 8, Moca

Acanthaceae (acanthus family)   October

Jacobinia umbrosa is a beautiful shrub, in the wilds or in the garden. In full bloom in the autumn, the shrub is covered with thick racemes of a brilliant, rich yellow. An excellent place to see the plants in cultivation is the garden of the Antigua Hotel, Antigua. The specimen shown here came from the coffee *finca* area, where these flowers were a spectacular sight in the wild. The shrub grows in Belts 7 and 8 and in damp woods of southern Mexico and Central America.

The plant was sent to England from Honduras and Mexico around 1840 and has been cultivated there ever since. It is grown outdoors in warm climates where there is sufficient humidty and also often under glass. It is said to be a great favorite of hummingbirds.

Kohleria elegans   Belt 10, Fuego Volcano

Gesneriaceae (gesneria family)   Autumn

The gesneria family is made up of about eighty genera, most species of which grow in tropical or subtropical regions. A few are small trees or shrubs; most of them are herbaceous. Achimenes and gloxinias belong to this family. All the flowers have irregular corollas, with at least one petal shaped like a little sack, or pouch.

Species of *Kohleria*, although attractive small shrubs, are not very well known. More familiar than *K. elegans* are two others from South America that have been cultivated: *K. amabilis* and *K. hirsuta*, both of which have larger flowers. *K. elegans* grows to thirty inches tall and has soft, hairy leaves and stems. The leaves grow up to six inches long, and the bright orange-red and yellow flowers make the roadsides gay toward the end of the rainy season. They are found on moist grassy slopes. This specimen was obtained on the road up Fuego Volcano.

These plants can be found in flower markets in Mexico and Central America. The flowers and leaves are used in preparations to cure stomach ailments, ulcers, and diarrhea.

The genus was named for Michael Kohler, a nineteenth-century teacher of natural history in Zurich, Switzerland.

Lobelia laxiflora (aretitos, chilpanxochitl)　　　　Belt 10, Antigua

Lobeliaceae (lobelia family)　　　　Spring

According to Bailey and Bailey's *Hortus Second*, lobelias are found in countries of moderate climates all around the world, including South Africa, Australia, Chile, Central America, Mexico and in the United States from Texas to Florida and even in Maine and Labrador.

Many species have been developed horticulturally for the garden, including the best known of all, the little sapphire-blue edging plant, *Lobelia erinus*.

L. laxiflora grows in all the temperate regions of Guatemala and is in bloom much of the year. It is not one of the most spectacular flowers but becomes like an old friend. *L. laxiflora* of Mexico has the name in common but looks slightly different as a herbarium specimen. However, there are very few duplications of identical species in Mexico and Guatemala, although as many genera grow in the almost identical subclimates as one would expect.

Lycaste virginalis alba (monja blanca—white nun)   Belt 5, Cobán

Orchidaceae (orchid family)   October

The beauty of this orchid is subtle and subdued but breathtaking. A faint glow of colors tints the pearly whiteness, and the "little white nun" is there behind her lectern.

Lycaste virginalis alba is a native of Alta Verapaz, where it grows in humid forests at elevations of four thousand to as high as seven thousand feet. The pink variety is well known to all Guatemalans and is more common than the white. Monja blanca, the national flower of Guatemala, has been widely popularized in the literature of the country and has been depicted on a Guatemalan postage stamp.

For many years this orchid was known to horticulturists in that part of the world and elsewhere by the botanical name *L. skinneri*. In their recent work on the orchids of Guatemala Ames and Correll found it necessary to change the name to *L. virginalis alba*. For sentimental reasons the change in designation is regrettable, since Skinner did so much to bring Guatemalan plants to the attention of horticulturists.

In the same area where these species of *Lycaste* grow are many other orchids that like the cooler weather.

"Species," or wild-orchid, plants, also known as "botanicals," are becoming more valuable and popular. A great many people all over the world now collect orchids, enjoying the novelty of creating their own hybrids. Not long ago a great many small genera were virtually unknown to the public; now they are common in greenhouses and shows.

Malvaviscus arboreus (tulipán)   Belt 11, Agua Volcano

Malvaceae (mallow family)   Summer

There are perhaps one thousand species in this family, a large number of which originally came from Mexico and Central America. The species shown here is common to Belts 10 and 11 in Guatemala.

Many showy species of this genus—herbs as well as a few shrubs and trees—are familiar garden plants. Included are the hollyhock, members of the *Sidalacea* genus, abutilon, hibiscus, and cotton (*Gossypium*).

A dozen or so species from tropical America are in use today in the horti-cultural world. They resemble fuchsias in many ways—the petals never open wide, and the stamens protrude. The flowers often nod, and they are invariably some shade of red. They grow in semishaded spots in rich soil with plenty of moisture. These lovely shrubs bloom all year, generally in areas shaded by trees and with fuchsias and ferns. The hibiscus thrives in warmer surroundings and is popular in sunny gardens in Guatemala City, but it is cultivated more widely in the southern United States and Hawaii.

Odontoglossum grande (boca de tigre)   Belt 5, Cobán
Orchidaceae (orchid family)   October

In their work on the orchids of Guatemala, Ames and Correll state that *Odontoglossum grande* is the largest flowered orchid ever seen in Guatemala, adding that it has been used extensively in horticulture. It is, indeed, showy and fairly abundant on the sunny slopes of the volcanoes in the central part of the country, as well as on tall trees in ravines near Antigua. Like *Laelia superbiens* and many species of *Oncidium*, as well as several other highland orchids, *O. grande* is often brought into the markets for sale or sold from house to house. Unfortunately for those who wish to grow orchids in their gardens, the Indians have a habit of bringing sprays of flowers with rootless pseudobulbs, rather than clumps that would be very easy to grow. It takes time to develop a good clump with a single new bulb a year, even when a few roots are attached.

This orchid has a distinctive common name—flor de tigre (tiger flower). It likes the climate of those regions of moderate rainfall and coolness, between five thousand and eight thousand feet in elevation. Ames and Correll indicate that Skinner brought this orchid to the attention of botanists in the early part of the nineteenth century.—W. P.

Oncidium splendidum (dancing ladies)　　　　　Belt 7, Escuintla

Orchidaceae (orchid family)　　　　　　　　　　March

Ames and Correll write: "This species, *splendidum*, is the largest and the showiest *Oncidium* found in Guatemala, where it is apparently endemic and extremely rare." They are right on all points. It certainly cannot be called a common wayside plant, but it is included here because of its beauty. I know of only two or three places where it has been seen wild. One is near Lake Atescatempa, near the El Salvador border, where the botanist Margaret Lewis discovered it about 1945. She wisely refused to tell anyone exactly where she found it because she had seen only a few specimens and feared that commercial orchid collectors would clean out the little colony.

The species is known to be more abundant in Honduras at elevations of around three thousand feet. It was brought into the markets of Tegucigalpa in flower at fifty cents a *masa*. I bought some once, assuming that a *masa* was a small clump. I found that a *masa* was five clumps and that we had bought two sackfuls.

Oncidium splendidum has been grown in Antigua for some years, but the climate there is apparently too cold, since its growth is exceedingly slow. The plant is very much like *O. cavendishianum* in appearance, but the large, heavy leaves of the latter, rather than being bright green, have a purplish tone. To my knowledge the panicles do not carry as many flowers as those of *cavendishianum*. I bought this plant near Escuintla. I found other plants there that I hoped were *O. splendidum*, but the leaves never grew very large, and when a spray of flowers opened, the plants proved to be *O. microchilum*.

There are about 1,500 species of orchids in Guatemala, many of which are small and of great interest to collectors. Several are well known, some of them occurring from Mexico south through Venezuela. One fascinating genus is *Pleurothallis*, about 90 species of which are indigenous to Guatemala. Among the many interesting plants are approximately 70 species of *Maxilleria*, 70 of *Oncidium*, and about 175 of *Epidendrum*.—W. P.

Petrea volubilis (bejuco—frog skin) Belt 6, Motagua

Verbenaceae (verbena family) February

Some nonbotanical visitors to Guatemala have suggested that the inflorescence of this beautiful half-climbing shrub, which can be seen wild in the Motagua Valley and more commonly as a cultivated plant, resembles a cross between a lilac and a wisteria. This species differs principally in its more-or-less climbing habit from *Petrea arborea*, which is a true shrub. Both are equally beautiful and are valuable additions to any landscape or garden since they bloom frequently and remain in flower for a long time. When the *petrea* species, are in full bloom, with their drooping panicles of light-blue flowers (whose starlike sepals remain on the plant for a long time), they are a glorious sight. The pansylike corollas fall off intact, and the ground underfoot becomes bright blue.

In the Motagua Valley (Belt 6) near El Rancho, the plant is simply called bejuco (climbing plant). Another local name is frog skin, from the texture of the leaf. This is strictly a Central American species, and is not mentioned by Standley as occurring in Mexico. In Guatemala it is "frequent in thickets and dry forests of the Pacific *tierra caliente*," Belts 1, 6, and 7.

The plant was introduced into England in 1731, and is also grown in warmer areas, such as Florida, where it is known as purple wreath. In other regions of warm climate it is called sandpaper vine.—W. P.

Plumeria rubra (cacaloxochitl, el palo de la

Cruz, frangipani) Belt 6, Motagua

Apocynaceae (dogbane family) June

The traveler who passes through the Motagua Valley during the spring or early summer will see, rising from among the dry shrubs and cactus along the roadside or the railway, occasional small trees surmounted by large clusters of glistening white flowers. Early in the season the trees are often leafless, but later their thick branches bear sprays of large, stiff, lanceolate leaves. If the traveler inquires locally, he will be told that this tree is el palo de la Cruz (the tree of the Cross).

The tree is the native and typical form of *Plumeria*, known around the world as frangipani, and is among the beautiful plants that Central America has contributed to the horticultural world. It is now one of the popular and romantic flowers of the tropical areas, not only in the pure-white form but also in combinations of yellow, pink, and white and also rose and yellow. In the Hawaiian Islands a variety that rapidly grows new, dark, glossy leaves does not seem to be deciduous. It was recently developed in Singapore and is now widely grown in gardens. All the varieties are beautiful, and frangipani is so fragrant that it has been utilized as the base for expensive perfumes. The blooms are used in making leis in Hawaii.

Plumeria rubra shown here is from a very old cultivated plant in the plaza of the churchyard of Santa Catarina de Bobadilla on the outskirts of Antigua. The species is seen in the wild in Belt 6.

Frangipanis are easily propagated by means of large cuttings—branchlets up to an inch or more in diameter. They like a warm or even hot climate and withstand poor soil and a dry season, as anyone who sees the plants on the mountainsides of Guatemala will realize. They are sufficiently cold-resistant to be successful in southern Florida and in favored areas of southern California. Their growth is slow, but where they find congenial surroundings, as in parts of Hawaii, they bloom profusely year after year and are a delight of the home gardener.

PLATE 31

Portlandia platantha (anacahuitl)   Belt 10, Antigua

Rubiaceae (madder family)   June

We found this beautiful small tree in an old churchyard in Antigua. After diligent searching in libraries, I can find no reason to believe it to be native to Guatemala, but I think it so lovely that I have included it with the following significant comment from Bailey and Bailey's *Hortus Second*: "Shrubs or small trees from Mexico and West Indies . . . one species grown in American tropics . . . platantha . . . somewhat resembles a Solandra." The painting agrees with the illustration in *Curtis's Botanical Magazine* (1850, pl. 4534).

Portlandia is a genus of fifteen to twenty species, most of which are found in the West Indies. Many familiar plants of this family are listed with the *Rondeletia*.

Patrick Browne, an Irish naturalist and botanical explorer of Jamaica, named the tree for the Duchess of Portland (1715–85). *Portlandia* is also grown in Hawaii.—W. P.

Psittacanthus calyculatus (liga, matapalo)   Belt 6, Salamá

Loranthaceae (mistletoe family)   June

Driving along the road in the warm valleys of Guatemala in the autumn, one often sees a tree that appears to be producing many spikes of orange-red flowers. Sometimes they are so numerous that they almost cover the tree and give one the impression that the tree itself is in bloom. However, they are the inflorescences of a plant of the mistletoe family, a parasite that attacks trees of many species. Although this plant is interesting in itself, it is better known for the so-called *flores de palo* ("wood flowers") which it forms at the base of the parasite where it is attached to its host. These curious scars, as Standley and Steyermark have written, somewhat resemble conventional rosettes of architectural decorations left upon the woody host plant when the base of the mistletoe plant is pulled away. They are often kept in houses for decoration and are occasionally sold in tourist shops. They vary in size from a few inches to about a foot in diameter.

In Guatemala the name liga is given to many plants of this family, which are frequently troublesome pests on fruit trees and many other economically valuable and cultivated plants. They are difficult to remove without cutting away the branches to which they are attached.—W. P.

Quercus skinneri (ruhji) Belt 5, Cobán
Fagaceae (beech family) June

This remarkable oak tree, *Quercus skinneri*, found in Alta Verapaz in northern Guatemala, looks more like a large-leaved chestnut tree than an oak, with its lanceolate leaves, six inches or more in length, on which each lateral vein terminates in a short spine. It is a handsomely erect tree up to about fifty feet in height, and produces extraordinarily large acorns, as shown here.

In the Kekchi language of the Alta Verapaz, this tree is called Ruhji (pronounced "roo-he"). The acorns fall to the ground, where they lie until they rot (which is a long time, despite the wet climate of this part of Guatemala). Recalling that hogs are fattened upon the acorns of the cork oak in Portugal and Spain and those of other oaks in France and Germany, one wonders why these acorns are not used as feed for domestic animals. Christopher Hempstead, of Cobán, is now experimenting along this line. If hogs fed upon these acorns resulted in hams as delicious as those produced in southern Portugal and Spain, it would be a real achievement (especially since the country people of Guatemala do not have much meat).

This tree is another example of the many extremely interesting and in some cases important plants brought to the attention of botanists by Skinner, whose descendants still live in Guatemala. He arrived in Guatemala before many plants had been sent to Europe, and he apparently had a keen eye for interesting material.—W. P.

Robinsonella cordata

Malvaceae (mallow family)

Belt 11, Fuego Volcano

June

This small tree is an astonishing sight in full bloom along a forest road. The pale blue, contrasting with the surrounding greens and browns, has the shimmering lightness of silver. It grows over a widespread area but apparently not abundantly, for not many are acquainted with the tree. In southern Florida it is grown occasionally as an ornamental.

The leaves and flower form are typical of the family and show the feature common to all its flowers: the stamens grow together to about half their length and protrude stiffly from the corolla. Note the form of the next hibiscus you see, or the little species of *Sidalcea*.

Rondeletia cordata (bouquet de la reina)   Belt 11, San Rafael

Rubiaceae (madder family)   March

Years ago, before the construction of the highway between Guatemala City and Antigua, visitors went by stagecoach through Mixco and up a narrow, lovely canyon to San Rafael, where they lunched at a tiny inn set among pine, cypress, and oak trees surrounded with masses of pink and blue hydrangeas. A little farther on they crossed the divide at seventy-two hundred feet, dropped down the steep canyon of the Pensativo into the Valley of Panchoy—the site of the old capital, now Antigua.

After the end of the dry season many lovely flowers bloom in the canyon between Mixco and San Rafael. One of them sure to catch the eye of the traveler is the slender, graceful shrub that produced the fragrant flowers shown here. This species, known in Mexico as jungle queen, belongs to a noble family, the Rubiaceae, to which also belong such plants as Cape jasmine, gardenia, bouvardia, coffee, and chinchona (quinine), as well as plants producing various dyes and medicines.

Rondeletia cordata does not seem to be common in Guatemala; at least we have not observed it in many places. It is well known in Mexico.—W. P.

Salvia wagneriana
Labiatae (sage family)

Belt 11, Fuego Volcano
October

Apparently unknown to horticulture is this very tall, handsome *Salvia wagneriana*, which we found on the Fuego road on a hillside with morning sun and afternoon clouds. It grows four to five feet high, and its colors are very striking against green shrubbery. The pinkish-red flowers are large and as attractive in bud as when open, since they emerge from large bronze bracts, which fall as the flower opens.

Other species of *Salvia* are known throughout the warm areas of North America and Europe, especially in areas of intense sunlight. There are many handsome natives west of the Rocky Mountains, and some are used in their natural form in gardens. Others have been developed by hybridizing into brilliant, hardy horticultural plants.

The best-known species of the genus is *S. officinalis*, the herb of garden sage, Old World in origin, the leaves of which are used for food flavoring in many parts of the world.

PLATE 37

Sprekelia formosissima (amacayo)　　　　　　　　Belt 5, Cobán

Amarylliaceae (amaryllis family)　　　　　　　　June

Sprekelia formosissima must be called a wayside plant of Guatemala because it is commonly seen, at least in Alta Verapaz. There has, however, been some question about its Guatemalan origin. It is well known in Mexico, and Standley considered that it extended southward into Guatemala, which is probably true because it is so abundant around Cobán. In appearance the large flowers, four to five inches in length, are strikingly handsome and almost spectacular. They stand above the slender long leaves at a height of twelve to eighteen inches above the ground. Like many other amaryllis plants they are to be seen only in one rather short period of two or three months during the year. They are crimson or crimson-scarlet in color, very showy, and excellent for cutting.

　　The bulbs multiply rapidly in the ground, as do many other amaryllis bulbs, and propagation is not a tedious job. *S. formosissima* is offered by nurserymen in the United States (and presumably elsewhere) as a house plant or for cultivation in gardens in warmer regions. One or two varieties are available, differing in size and somewhat in color.—W. P.

Tabebuia donnell-smithii (matilisquate,

  palo blanco, primavera)   Belt 8, Moca

Bignoniaceae (bignonia family)   February

The two beautiful species of *Tabebuia* of Central America are outstanding in the glory of their blooming period—the latter part of the dry season. Both species form an umbrella of color above a long, straight trunk, and they reach high above their neighbors for sun. *Tabebuia pentaphylla* is the national flower of El Salvador and is there called macuelizo. Its flowers range from a lovely pink to rose. *T. donnell-smithii* has brilliant, clear-yellow flowers. It is sometimes called palo blanco because of its white bark, which enhances its attractiveness. It is common to the low and middle altitudes—the warm Belts 1, 7, and 8.

This tree is very useful as well as decorative, for it grows well under adverse conditions and propagates easily. Under the name gold tree it is now being planted around Honolulu as a part of the city's beautification program. Sometimes this tree is called primavera (spring) or the bignonia tree.

Another member of the Bignoniaceae is the ubiquitous jacaranda, a native of Brazil. Its individual flowers are not as large or lovely as those of *T. donnell-smithii*. *Tecoma stans* (Plate 41) is another member of the family, as are the crescentias (calabash trees) of the desert.

Most flowers are carefully wrapped, or much underdeveloped, in the bud, but the *Tabebuia* species are different: the petals are large and look as though they have been stuffed into a shell. When they burst open, they are a mass of dark wrinkles on the first day. This makes them impossible to paint until they are well established, and when they are at their peak of perfection, there are too many flowers to show distinctly. Thus I have had to leave out some to show the form of the individual blossoms.

Tagetes—four species

    (marigold, flor de muerto)   Belt 10, Antigua

Compositae (sunflower family)   October

Marigolds are among the commonest wayside plants in Guatemala in the autumn. Shown here are four of the smaller species. They grow in gay drifts and are a delight in the temperate Belt 10 and all the way from Costa Rica through the highlands of Mexico.

Standley discusses seven species in his *Flora of Costa Rica* but it is difficult to differentiate these small ones shown here. I believe that the two left-hand specimens are *Tagetes remotiflora* and *T. sororia*, but I am not sure of the others.

There has been a great controversy about where *Tagetes* originated. All the regions have cities that were ports of long-ago Spanish ships: eastern India, North Africa, Portugal, and Central America, as well as Spain. One reason for the confusion has been that in all these areas there are religious rites in which *Tagetes* is used. The German botanist Leonhard Fuchs, who named the genus in the sixteenth century, claimed that the place of origin was the New World. Two points that support this claim are the fact that species of *Tagetes* are shown in Maya carvings made in the great days of that civilization—that is, before A.D. 1000—and that *Tagetes* were given an Aztec name—cempoalxochitl.

These little flowers are used extensively in the Indians' rites on both All Saints' Day and All Souls' Day. The use of flowers is obviously a remnant of pre-Columbian ritual. Bouquets are hung in doors and out, and on All Souls' Day paths are made of petals from the cemeteries, so that the souls of the dead can find their homes.

Tecoma stans (tronadora)   Belt 6, Motagua

Bignoniaceae (bignonia family)   October

Every traveler in Belts 6 and 10 is certain to see this showy shrub (also called *Stenolobium stans*). Sometimes it becomes a small tree. The panicles of large, tubular, bright-yellow flowers and the long seed pods make this plant easy to recognize as a member of the Bignoniaceae family. *Tecoma stans* has become a popular plant in southern California gardens because it is remarkably easy to cultivate and is somewhat resistant to cold weather. In fact, it seems to be "native from Southern U.S.A. to Argentina," according to Bailey and Bailey's *Hortus Second*.

T. stans is not often planted by Guatemalans in their dooryards since it is so common in the wild. It is not seen in the rain-forest regions but is found throughout the country in drier regions at elevations between one thousand and five thousand feet. At the higher altitude the plant is abundant. It is very easily propagated by means of seed and has a long blooming period.

Other members of the Bignoniaceae are the *Crescentia*, catalpa, and the favorite jacaranda, which can be seen in the churchyard of almost every town.—W. P.

Tillandsia rodrigueziana (pata de gallo) Belt 11, Agua Volcano
Bromeliaceae (pineapple family) Spring

Bromeliads are among the most spectacular of flowering plants, and, being epiphytic, they are usually found dramatically high on branches of trees. There are about forty genera, all native to the New World. They cannot withstand frost or drought and grow best with plenty of light and heavy rains in season and with some humidity at all times. The species shown here is seen in Belts 8, 9, and 11.

The blooms of the *Tillandsia* often have local names reflecting colorful comparisons. The common name given here is Spanish for rooster's foot. Another name used rather loosely for spectacular blooms in this family is guacamaya, the name of the gaudy macaw. The same name is also given to the bird of paradise, *Strelizia reginae* (Musaceae), and the beautiful *Phyllocarpus sepentrionalis* (Leguminosae), mentioned by Steyermark.

As foliage plants members of the *Tillandsia* genus are excellent in pots for the house or for hanging baskets in the garden. They are popular nowadays as such, and some can even be found in variety stores all over the United States, but these specimens are tiny and require years to mature. Some plants grow to be one yard in diameter, with huge bands of vivid color.

There are many species and varieties of this genus. One is Spanish moss (old-man's-beard), which grows in the old oaks of the southern United States and of northern California. Another well-known bromeliad is *Ananas comosus*, the familiar pineapple. This plant is cultivated for commercial use, principally in Hawaii. The delicious pineapples of Guatemala are an important crop.

Tithonia longiradiata (giganton, tree sunflower)   Belt 10, Antigua

Compositae (sunflower family)   January

By the roadsides near Antigua in more or less the same kinds of locations as the cassias and wigandias (Belts 6, 7, and 10) will be found these large, sprawling shrubs. They bloom in the early months of the year and are spectacularly handsome. I have not seen any that look like trees, but the plants have a woody base, which botanically place them in that category. There is no doubt that they are sunflowers. The large, fully petaled flowers are augmented by heavy, velvety leaves, which twist and curl dramatically.

Tithonia longiradiata is found in the mountainous areas of Mexico through Guatemala and as far south as Costa Rica, where part of the year is dry but not too cold.

Flowers of the family Compositae, to which *Tithonia* belongs, are actually an assemblage of two kinds of flowers, from four or five to twelve or more of each variety. In the center is a grouping of ray flowers, each one with stamens and pistil and usually four petals. Outside this group is a ring of flowers that look much the same, but one petal is extended out, and all these disk flowers act like the spokes of a wheel. What we see at a casual glance is usually the latter. On ripening, each little flower becomes its own seed carrier.

Triplaris melaenodendron (mulato) Belt 7, Escuintla

Polygonaceae (buckwheat family) January

With luck, during the early months of the year one may see specimens of this tree in all their glory when they are in fruit. They have attractive sprays of seed cases with "wings" rather like those of maples. The species is deciduous and for at least part of the year has no conspicuous characteristics. When the large leaves grow and the small whitish flowers are in bloom, the trees are recognizable. Any doubt can be resolved by breaking off a branch: a *Triplaris* branch is hollow and full of voracious ants. Another characteristic is that the stumps send up new sprouts, which form a little thicket. The common name mulato derives from the mottled bark.

According to Standley and Steyermark: "The tree is abundant in many parts of the Pacific Plains and often affords wide displays of color in late January and February. It is one of the most characteristic species of the Pacific Coast of all Central America" (*Flora of Guatemala*).

We found this specimen near Escuintla. The panicle shown is another of those where in the painting elimination of some details was necessary; the specimen was thick with little wings.

There are several related, but largely undifferentiated, species up and down the Pacific Coast. They are planted for ornamentals throughout the tropical regions of the Western Hemisphere. Other genera belonging to the Polygonaceae family include *Chorizanthe, Eriogonum,* and *Antignon.*

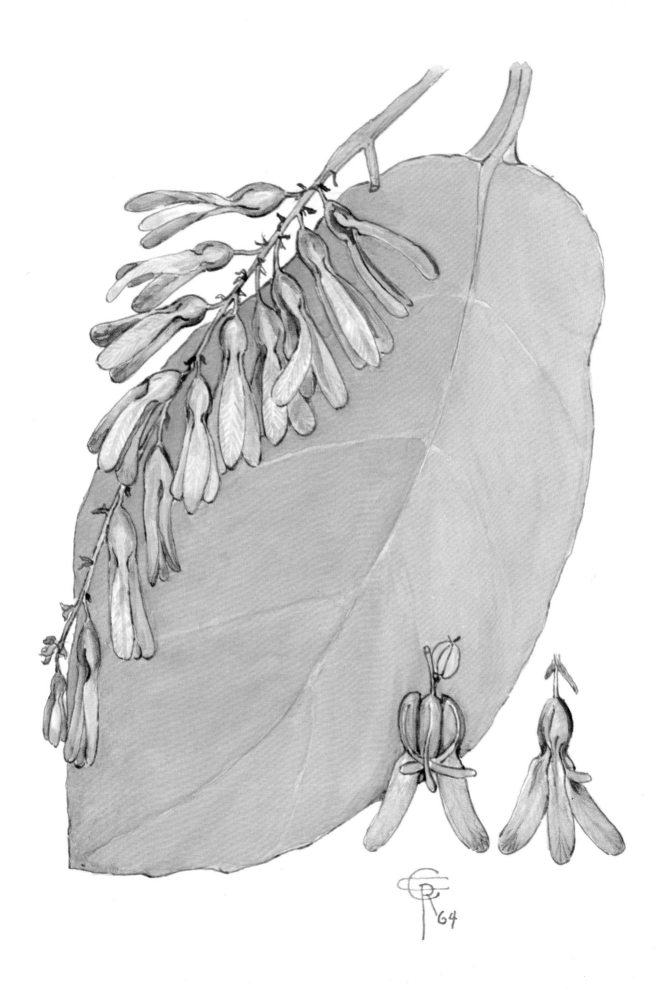

Weldenia candida   Belt 9, María Tecum

Commelinaceae (spiderwort family)   Summer

Plants of the Spiderwort family are generously scattered in frost-free regions where there is adequate moisture. Commonly known as wandering Jew, they are found in Guatemala among the lush growth of canyons in the cloud forests, as well as in Belts 10 to 12. These plants are easily recognized by the overlarge bracts from which relatively small flowers (generally blue or white) open but for a day in sequence. *Tradescantia* is the best known genus in horticulture.

This species of *Weldenia* was found by the side of the road called María Tecum, a pass through mountains between Mazatenango and Quetzaltenango, at about eight thousand feet. The region is made up of tundra and old cypress trees, with undergrowth below the smaller trees that resembled huckleberry (the family Ericaceae). On outcropping rocks are lichens and mosses, and a familiar-looking species of *Echeveria* can be seen among other rock dwellers. Covering the ground in the shade is a layer of sphagnum moss six to twelve inches deep. Growing up through it are two kinds of flowers, a little geranium and the *Weldenia candida*. The painting illustrates how the root emerges from the ground and extends itself through the moss to find the light. This plant is so sensitive that it half-closed in the shade I briefly made. There is a lovely little bouquet of flowers in each plant.

Wigandia kunthii (chocon)	Belt 6, Motagua

Hydrophyllaceae (waterleaf family)	February

When a new road is built in the hills of Guatemala that necessitates deep cuts, one of the first plants to make its appearance on the steep slopes is *Wigandia kunthii*. It is also likely to make its uninvited appearance on patio walls, if there are rough spots on which it can gain a foothold. If not removed, the plant in time will play havoc with the wall, for it becomes a very large, coarse shrub (the stems are sometimes thick enough to be used for firewood). The leaves are large, coarse, and, as Tracy Hubbard wrote in Bailey's *Standard Cyclopedia of Horticulture*, "not only hispid [bristly] but *very* hispid." The stiff, short hairs that cover them sting like nettles. The plant is also common to Belt 10.

W. kunthii has been used as a foliage plant for subtropical bedding in some regions, notably southern California, but one cannot say that it is really popular. I have never seen it cultivated in Guatemalan gardens, but woe betide the homeowner who tries to pull one off the top of his wall, if his cook belongs to the old school and makes tamales from time to time. Tamales do not have the right aroma if one does not place several chocon leaves on top of a potful before she puts on the lid. I wonder how the Mexicans have been able to make such fine tamales down through the centuries without chocon leaves.—W. P.

Yucca elephantipes (flor de isote)   Belt 10, Antigua

Liliaceae (lily family)   February

When this common plant attains an age of fifty years or more, it well deserves the name elephant's foot, because the base of the plant has become very thick and remarkable in appearance. It can be seen all over Guatemala, primarily in the middle elevations and the not-too-wet areas, planted as tall hedges or along steep roadsides to prevent erosion. The common name is isote, and in the markets are often seen what are known as flores de isote. The flowers are used as human food, sometimes served with pickled beets, which give the flowers a lovely color. They are best served fried in batter and are considered a delicacy. They are popular and no doubt contain vitamins.

It is interesting to speculate upon the matter of vitamins. How did the Indians discover, perhaps centuries ago, that they needed them? Of course, they did not call them by name, but they ate a number of leafy plants which have no particular flavor but are valuable sources of vitamins and minerals. Perhaps the best example is a plant grown in Mexico, which Robert Harris, of the Massachusetts Institute of Technology, brought to the attention of food experts some years ago. It is in fact a variety of wheat that grows abundantly on the plateau of Mexico and is cooked in soups and similar dishes. Harris found that this "weed" was a remarkably valuable addition to the Mexican diet, which, in the highlands, is otherwise lacking in certain minerals and vitamins.—W. P.

   PLATE 47

Zephyranthes brevipes (atamasco, flor de mayo)   Belt 10, Iximché

Amaryllidaceae (amaryllis family)   June

With the onset of the rainy season in the Guatemalan Highlands, this interesting little amaryllis springs up in open grassy places. It belongs to a genus of about forty species in tropical and temperate parts of America. One of them, *Zephyranthes brevipes*, is the atamasco, or zephyr, lily of the United States.

The specimen shown here grows in an area with a romantic background, for it was collected in the plaza of Iximché, the ruins of the Cakchiquel capital near Tecpán, at an elevation of about seventy-five hundred feet. The region is one of pine and oak forests, many of which were cut, doubtless in early times, to make way for cornfields. Iximché, like Utatlán and Zaculeu, was not founded at an early date, and was in its prime at the time the conquistador Pedro de Alvarado marched down from Mexico to conquer and colonize Guatemala. He met a friendly reception at the hands of the Cakchiquel people, and was so well cared for at the start that he wrote home that "we could not have been better off in our father's house." After a few months, however, the Indians had been so badly treated that they abandoned the region, and Alvarado eventually made his capital at what is now called Ciudad Vieja (Santiago), in the lovely Valley of Panchoy, which lies at the foot of Agua Volcano.

While there are no interesting sculptures at Iximché, there are the remains of many sturdy stone platforms and temple mounds, on which an outstanding work of restoration has been achieved by the Swiss archaeologist Georges Guillemin. There are not as many wild flowers here as may be found in certain other parts of Guatemala, but visitors interested in the background of Guatemala should not fail to include Iximché in their travels.

Zinnia elegans (carolina)   Belt 6, Salamá

Compositae (sunflower family)   June

The zinnias, like the marigolds, have made a great contribution to North American culture. Especially in the valley of Rabinal, some distance north of Guatemala City, the species shown here is abundant along the roadsides. Locally it is called carolina.

This is one of the plants that causes us to wonder what the Indians grew in their gardens before the arrival of Columbus. Today roses and many other flowering plants dear to Spanish hearts are commonly seen in the dooryards of the Indians. (I do not mean, of course, that Columbus brought them personally, though he did bring seeds of various fruit trees, cereals, and other edible crops.)

The Guatemalan Indians are flower lovers, and there is scarcely a house, no matter how small, without a few flowering plants around the door. Among the common ones are some of the native species, such as the zinnias, the marigolds, and the tree dahlias. The native canna is also popular as a garden plant at middle elevations. Of introduced ornamentals, it is natural that the rose should be the most popular. But there are also many geraniums, calla lilies, gladioli, watsonias, carnations, and other cosmopolitan ornamentals.—W. P.

Fuego Bouquet   October
                                       Belt 11, Fuevo Volcano

Among the small flowers that line the dirt roads of Guatemala are many that will seem similar to those in the gardens of the United States. This bouquet is a small handful picked at random one day. About half of the plants are known to horticulture; the others are gay little wildflowers.

From left to right they are *Quamoclit grandiflora* (Convolvulaceae), *Salvia urica* (Labiatae), *Lamoureuxia multifida* (Scrophulariaceae), *Bidens pilosa* (Compositae), *Cuphea pinetorum* (Lythraceae), *Calceolaria mexicana* (Scrophulariaceae), *Odontonema strictum* (Acanthaceae), *Bouvardia ternifolia* (Rubiaceae), and *Crusea calocephala* (Rubiaceae).

Index